Ladies of
Blaenwern

The Ladies of Blaenwern

The story of The Dorian Trio and the Llanarth Welsh Cob Stud

TELERI BEVAN

First impression: 2010
Second impression: 2011

Cover design: Y Lolfa

ISBN: 978 184771 263 9

Published, printed and bound in Wales
by Y Lolfa Cyf., Talybont, Ceredigion SY24 5HE
website www.ylolfa.com
e-mail ylolfa@ylolfa.com
tel 01970 832 304
fax 832 782

PRELUDE

M USIC AND FARMING ARE the twin themes which connect the lives of the three ladies of Blaenwern known as The Dorian Trio and owners of the Llanarth stud of Welsh cobs. Pauline Taylor and Enid Lewis, both professional musicians, began their concert work as friends at Birmingham University during the First World War and later they teamed up with Barbara Saunders Davies in the mid-1930s. The friendships evolved into unique business partnerships. Successful, remarkable and notable, they overcame prejudice during the years when as spinsters – the three never married – they were often the cause of ridicule and derision especially in the Ceredigion cob world. But these were not typical women. They were extraordinary; they possessed great energy, organisational skills and found various solutions to solve problems. It is a stirring story.

Generations of children who were brought up in Wales in the 1930s, 40s and 50s knew of The Dorian Trio. I was a rather unwilling pupil at Ardwyn Grammar School and endured those educational concerts in the school hall. As a teenager I was bored witless with the sound of chamber music, but I never forgot the joy on the faces of the three musicians and their total commitment to communicate that joy to a difficult young audience. Then, much later in the 1950s and 60s, I heard that the ladies had established the Llanarth stud and I heard of their success as breeders of all native Welsh breeds – cobs and corgis, Welsh Black cattle and Welsh pigs. I met Pauline Taylor many times – she was the spokesperson for almost all their activities. Slightly eccentric but an enthusiast to her fingertips, she had a deep love for Wales. Once they had earned the respect of the cob fraternity, they were regarded as visionaries way before their time, as they went about transforming old attitudes and practices. The stud became supremely successful.

5

Barbara Saunders Davies, after seventeen years breeding cobs, left Blaenwern to become the librarian at the Rudolph Steiner Centre in London before retiring to her county of birth, the Pembrokeshire Preseli hills. When Enid Lewis, who had become Professor of Piano at the Guildhall School of Music, retired, she invested her savings to buy the Blaenwern estate and she and Pauline formed a farming partnership until they died in the early 1980s.

But unfortunately, old age brought a tragic ending to the story, with the dismantling of the farm and stud by the University College of Wales, Aberystwyth, who had been gifted and bequeathed the estate and farming enterprise. Many will remember the acute anger and disappointment at the final sale, the dispersal of the Llanarth stud and the press headlines and the television programmes. Pauline and Enid died of broken hearts.

*   *   *   *

Coda: Early maps often have the place-name Llanarth in a different form; it is frequently seen as Llannarth. The *Gazetteer of Welsh Place-Names* places the double 'n' as the correct spelling, but common usage today and for many, many years has been Llanarth. It is the prefix used for the successful Llanarth stud, and for that reason I have used that form in this book.

ACKNOWLEDGEMENTS

I AM DEEPLY GRATEFUL to everyone who gave of their time to contribute to my knowledge of cob breeding and the lives of the Ladies of Blaenwern: Len Bigley, the trusted Llanarth stud manager, Wynne Davies, breeder and historian, Anne Wheatcroft, William Lloyd, Ifor Lloyd, Dai Jones, Myrfyn Jones, Professor Desmond Hayes, Gwawr Owen, Carole Knowles-Pfeiffer, Nim de Bruyne, Zena Lockett, Kathryn Thomas CVO and the late Professor Anthony Bradshaw and Anne Fowler. The Welsh Pony and Cob Society, the Pembrokeshire Archive, the National Library of Wales and the National Screen and Sound Archive of Wales.

And a special 'thank you' to Sterling Asset Management, in particular to its Principal, Gerallt Davies, for generously contributing to the cost of publication.

I invited two well-known stalwarts of agriculture and rural life in Ceredigion to write their observations as an introduction to *The Ladies of Blaenwern*: Dai Jones, Llanilar, is the President of the Royal Welsh Agricultural Society for the year 2010 and William Lloyd of the Geler stud recently relinquished his role as President of the Welsh Pony and Cob Society in March 2010. It is a remarkable coincidence that both have their family roots in the small parish of Llangwyryfon, and that the notable success of both societies during their tenure is testimony to their enthusiasm, knowledge and leadership.

Teleri Bevan

FOREWORD

IT IS WONDERFUL TO receive another book from Teleri Bevan about people who have contributed so much to the countryside communities of Wales. Teleri was a major influence in the production of countryside and farming programmes in the early years of BBC Wales radio and television. There is no doubt that pioneers such as Teleri are to be thanked for the commitment of broadcasters to the Royal Welsh Show of today. The television coverage that countryside issues enjoy in Wales has become the envy of all the Celtic nations, especially the hours of broadcasting on S4C from the annual four day Royal Welsh Show in July and the Winter Fair, just before Christmas.

Teleri comes from the best of Welsh families. Her father, the late Dr Richard Phillips, was a strong influence in the early years of University College of Wales, Aberystwyth, and at the Welsh Plant Breeding Station in Gogerddan. Like many members of the Young Farmers Clubs, I owe so much to Dr Phillips for teaching young people like myself to do our best for the farming and countryside of Wales and the importance of our rural life and traditions.

I look forward very much to this latest book, *The Ladies of Blaenwern*. I wonder how many of you remember the lovely ladies coming to our schools and giving us great joy with their music making. And of course, we are grateful to them for their sterling work in breeding our great Welsh cobs and for the prefix Llanarth which carries on today.

Teleri often returns to the family home at Argoed, Llangwyryfon, where her bubbling personality is something to treasure.

Dai Jones MBE
President of the Royal Welsh Agricultural Society, 2010

I FEEL VERY PRIVILEGED and humbled to be asked to write this foreword to this long-awaited publication, *The Ladies of Blaenwern*, which has been uniquely written and chronicled by Teleri Bevan.

The partnership formed by Pauline Taylor, Barbara Saunders Davies and Enid Lewis at Blaenwern which established the Llanarth Welsh cob stud, meant that things would never be quite the same again in the Welsh cob world. The personality, charisma and strength of character endeared the ladies to all who came into contact with them, but in time, this would become secondary to their successes and results, especially in the Section D Welsh Cobs.

The Section D of the Welsh Cob Stud Book was at a low ebb and the future looked bleak after the Second World War. But, by the end of the 1940s, there were signs of a recovery of interest. The ladies set up their farming enterprise at Blaenwern and by the early 1960s, the Llanarth stud was firmly established. Their involvement in cob breeding resulted in a marked increase in activity at shows and sales, which attracted support and participation from Wales, the UK and beyond.

In time, the ladies at Blaenwern instigated the Welsh cob sale at Llanarth where their own cobs and cobs from other studs were sold. The ladies strongly believed in all types of Welsh breeds and felt that breeders did not promote themselves beyond the local area. Their first sale was held at Blaenwern on the 17th of October 1964. This was a new and unexpected venture and Welsh breeders experienced a whole new world that they had previously not experienced.

I vaguely remember that occasion from the report and photographs that appeared in the Welsh-language weekly, *Y Cymro*, showing the Cardiganshire breeders and others seated in a circle of straw bales surrounding a roped ring. Little did anyone realise that history was being made that day. The sales became a phenomenal success story, growing beyond the wildest dreams of the Cardiganshire cob world.

I believe the courage, timing and foresight of the ladies of Blaenwern in staging of that first Llanarth sale and subsequent sales, coupled with their unselfishness in providing an opportunity for all breeders to participate, has proved to be a major factor is the rise and success of the Welsh cob breed. This is the lasting memory – an epitaph that the three ladies would have wished for.

William Lloyd
President of the Welsh Pony and Cob Society, 2009

CHAPTER ONE

IT IS ALWAYS DANGEROUS to write about real people's lives. There is always someone who will tell you, 'Of course she was quite mad.' If madness is a necessary condition to be different, to be resilient and remarkable, energetic and courageous, then the three ladies featured in this book qualify. They were born in the Victorian age; they thrived on their work ethic, academic and artistic, practical and pragmatic. Their lives spanned most of the twentieth century. They didn't follow role models, they didn't set out to change the world or to follow pioneering women of the past, but they did become single-minded achievers.

Enid and Pauline met in 1917 when they were studying music at Birmingham University. Pauline, a cellist and a native of the city was about to graduate in the subject. Enid, five years older, who was already a graduate of University College of Wales, Cardiff, had come to Birmingham to enhance her skills as a pianist. They became firm friends, they played at concerts together, and for the next fifty years their livelihood came from music making and lecturing, later becoming known to generations of children as The Dorian Trio. In 1936 they met Barbara Saunders Davies at a concert on the border between Cardiganshire and Pembrokeshire, and she was to play an integral part, not only in their concert work, but also as a knowledgeable geneticist when they began farming and establishing their Welsh cob enterprise at Blaenwern in the parish of Llanarth, near the shores of Cardigan Bay.

Pauline Taylor was one of five children – two boys and three girls – born to academic parents in Birmingham. Her father, John W Taylor, was a revered consultant gynaecologist and Professor at the city's university. Her mother, Florence, was a talented botanist and teacher. They had met at her home in London in 1868 when Florence was but a child, and John Taylor, on the threshold of a medical career, had come to be articled to her father, a general practitioner. His medical career flourished and Florence – herself a brilliant student – was not allowed to graduate as was the rule in those days. She had to be content with a first in part one Botany at Newnham College, Cambridge.

Twenty years after their first meeting they met again; he was immediately captivated by her beauty, maturity and intelligence and he asked for her hand in marriage. She was thirty-three and he forty-eight. Their wedding took place in 1889, in the year that John was appointed Professor of Gynaecology at Birmingham University. They were happy, being Victorians who believed in large families and five children were born in fairly quick succession. Without doubt it was Florence who was the main focus in the family during her children's formative years because, for a few years, she and her husband spent a great deal of time apart. Today their descendants are unable to explain why, but publicly at the time, the long-lasting reason for this unusual lifestyle was that John Taylor went to live with his sister where he could, 'better attend to his patients'. The children were brought up by their mother at the family home, Island Cottage, Northfield, but little is known of the effect that the two parents living apart had on the children. No-one talked about it, the Victorian stiff upper lip was kept appropriately stiffened. They just got on with life, but Pauline's regard for her parents, particularly her father, had a profound influence on her. A wise counsellor and a born peacemaker with a wonderful sense of humour, he never seemed to take a stereotyped view of anything or to leave a subject as he found it. He was a literary man, his volume, *The Coming of the Saints*, an interpretation of the legends of Mary, Martha, Lazarus and Joseph of Arimathea, had been published and he was much sought after to give lectures and to explore in essays themes on social and moral issues. At the request of the Bishop of London he gave a courageous address in 1904 on 'The Diminishing Birth Rate' saying, 'I have always held that the great ideal of the large and cultured family, where plain living, high thinking and holy aspiration are the three great features of the upbringing, is the very highest ideal of civilisation.'

In Victorian England their day-to-day living arrangements may have surprised their friends but both parents wrote of their marriage as being strong and happy. Florence assisted her husband to produce his medical and poetry publications. His book of poetry, *The Doorkeeper*, was well received when it was first published in 1910, shortly before his death of chronic heart disease at the age of fifty-nine. Florence could have had a brilliant career as a scientist but chose instead to teach her children at home until they reached high school age.

'She was a beautiful woman of intriguing character,' her daughter Anne wrote of her in a memoir, 'Sharp as a needle, but wayward and fanciful – gentle, but severe and decided – dreamy and practical, not by nature a good manager – sweet and loving – dynamic in energy and above all, of a steadfast unquenchable faith and courage.'

A crippling accident in her forties confined Florence to a wheelchair but with typical determination and courage she coped, as she did when faced with other distressing events in quick succession. Her husband John died in 1910 following a short illness, she nursed her youngest son back to health when he suffered a severe accident, and her other son joined the army and was sent to the front line at the outbreak of war in 1914.

Pauline's formative years were also coloured by the concerns and worries of these events, but it was clear from an early age that her parents believed she was destined to become a musician. It was an interest which she inherited from her father, an accomplished violinist and organist who derived great pleasure practising in his surgery, calming the nerves of anxious patients sitting in the adjoining waiting room. Both parents were taken aback when Pauline unexpectedly announced at the age of eight that she would like to be a horse breeder when she grew up, but she was well into her forties before she realised that dream. Unexpected thoughts were part and parcel of Pauline's character and her sister's description of her mother could well have applied to her.

Pauline put her hopes and dreams to one side, obeyed her parents bidding and continued her musical studies at Birmingham University, perfecting the skill of playing the cello. Three years later she graduated, and joined the chamber orchestra and teamed up with Enid Lewis to give concerts to soldiers returning home to recuperate from the horrors of the Great War. Many years later she would recall those times with the comment, 'It seemed the most purposeful thing to do'.

Enid Lewis was single minded and seems to have had a high proportion of musical genes in her make-up because her determination to be a musician became apparent from a very early age. An only child, she was born and brought up in Pontypridd, the gateway town to the Rhondda valleys, the heartland of the coal mining industry in south Wales at the turn of the century. Her father, John Lewis, the eldest of four brothers, was a native of the town and a prominent local businessman who developed a prosperous hay and corn business in a mill near the Merthyr-to-Cardiff canal. Strategically placed, the business provided feed for hard-working pit ponies and horses pulling barges loaded with coal along the canal between the coalfield and Cardiff docks. He built an impressive Victorian double bay house on a hill above the town appropriately called Tophill and it was here that Enid spent her early years, in comfort and given every opportunity to excel.

She inherited her musical talents from her mother, Sadie Thomas, who was apparently the first woman in Wales to earn the distinction of the diploma LRAM. She taught private pupils, she organised chamber music concerts, she was a founder member of the Welsh Folk Song Society and of the Pontypridd Cymmrodorion Society. Not content with being a mere 'good wife' and teacher, Sadie is remembered as a formidable, outspoken, independent woman, a devout Unitarian, public spirited, a teetotaller and an ardent suffragist who often attended protests and marches in London. 'She got things done,' Kathrin Thomas, now the Lord Lieutenant of Mid Glamorgan told me, as she recalled members of the family discussing the strength of Sadie's personality, 'You never, never, ever argued with Aunt Sadie.'

She was an ambitious mother and keen that her daughter should make the most of her talents and be proud of her gifts. From an early age she recognised that there was a pianist of rare promise in her daughter. Enid was a willing pupil with an aptitude for learning. She sailed through her schooldays and the matriculation examination at Pontypridd Girls' Grammar School before entering the music department at University College of Wales, Cardiff. Both her parents were devout Unitarians and it was through links with the church that, following her graduation, she spent two years as a student in Budapest where many virtuoso pianists gave up their time to teach at the Franz Liszt Royal Academy of Music. These were the formative years in which Enid became confident, meeting people, experiencing another culture and developing her technique and interpretative skills as a pianist.

She came home as the threat of war gathered momentum in order to consider options to further her career. During that time with her parents in south Wales, her father persuaded her to learn to play golf and she became so good at it that she was selected to represent the Radyr Golf Club ladies team at a team championship at Aberdyfi in north Wales. Golf was a diversion from news of the carnage on the battlegrounds of Belgium and France which affected her deeply, and she offered herself as a VAD (Voluntary Aid Organisation) nurse, which trained women in many skills, from cooks and clerks to ambulance drivers and fundraisers. She returned to Hungary for a short while – a country and people she had come to love – as a VAD volunteer, but it was under the auspices of the YMCA that she decided she could best support the war effort by giving concerts for soldiers at venues in the Midlands. It was at this time she teamed up with Pauline Taylor.

*   *   *   *

Pauline was already working as a freelance musician and Enid was keen to finance further studies and to perfect her playing. The Professor of Music at Birmingham University, Dr Granville Bantock, who had succeeded Edward Elgar in the post, was revered as a prolific composer, conductor and teacher and it was he who suggested that Enid should seek further guidance from the Matthay School of Piano. The school, based in London, was founded by Tobias Matthay, who had gathered around him concert pianists such as Myra Hess to pass on their experience and knowledge to young students. Matthay was considered the greatest pedagogue of all time and pupils received advice on artistic and a wide range of complex technical matters – from the laws of pedalling to the forearm rotation principle. His main maxim to parents, pupils and students was, 'Never touch the piano without trying to make music.' Enid needed little encouragement.

Pauline was by no means a reluctant musician – she found immense satisfaction and pleasure communicating with audiences through music – but the recurrent dream of her formative years was the vivid notion that one day she would own horses and live in the countryside. She had spent all her school summer holidays on a farm near Stratford-upon-Avon:

> I would take my cello to practise, and there I found horses and dreamed wild dreams. My favourite was a gelding called Jolly, I think he was a Welsh cob but I didn't know that then. I was so in love with

horses that I stuck a strand of hair from the mane of each horse on a sheet of paper and put the name of each horse underneath. They were put in order of love, and on top was Jolly – the gelding I worshipped.

That statement seems to reflect Pauline's attitude throughout her life. Her mother, an independent person had once written of her hopes for her children, 'I wish my children, girls and boys to be independent, so that each member of the family should have scope for full development.'

Pauline was practical, clever and forthright with an enormous zest for turning new thoughts and ideas into positive realities but, in later life, she invariably sought help from others if her ideas needed financial backing. Her partnership with Enid flourished, not only because they had both developed empathy as musicians, but also because they had a deep love of the countryside.

Physically and in personality they were very different. They were opposites but not obstructive. They could contradict each other, but they were not rivals. Enid was short with an open, warm and gentle face and that gentleness was reflected in her character. She liked people, especially those with artistic interests and, as a performer, she was at home on the concert platform communicating her own love of music with the audience. It was teaching young people that gave her the most pleasure. A sympathetic listener, her own enjoyment would be reflected in her gracious good humour and her smiles of encouragement.

There was a restless, emotional quality about Pauline. She was tall, her hair turned white at a young age, her blue eyes, always watchful, surveying in an instant everything around her. Her eyes also conveyed impatience, a pent-up energy and she never seemed to take a conventional view on any topic; she was occasionally argumentative and forthright, quixotic and inspirational.

Enid was a steadying influence. Her roots were in

Cardiganshire and as a child she would spend summer holidays there with her grandparents on a farm, Blaenwern – her mother's birthplace – in the parish of Llanarth. It was a place she returned to as the proud owner some fifty years later, but in those early years she and Pauline would often escape to the peace and tranquillity of the Cardiganshire countryside around the attractive Cardigan Bay seaside village of New Quay and the parish of Llanarth. Her grandmother was still alive and Enid would return regularly to stay with her.

The link with rural Wales grew ever stronger as the 1920s dawned. The decision to concentrate on chamber music and to form The Dorian Trio brought new opportunities and cemented her friendship and professional relationship with Pauline. They continued to freelance as individual performers but for The Dorian Trio engagements, they would call on violinist, Kathleen Washbourne. Their recitals drew encouraging notices in the press and their reputation grew, *The Sunday Times* critic wrote, 'The Dorian Trio think and play with refinement and consideration for one another and the music.'

When Pauline and Enid heard that Dr Walford Davies, the organist and composer, had been appointed to the Chair of Music at the University College of Wales, Aberystwyth in 1919, they hastened to Cardiganshire. He had often collaborated with Dr Bantock at festivals and concerts and both had aspirations to encourage music in schools. Bantock spent a great deal of time at his second home in north Wales and he made his views on music education in schools known in no uncertain terms, 'England is still one of the most backward musical nations. In Wales I saw great promise, provided they continue to develop their national love of music and encouraged it in their schools.'

Walford Davies had taken up a similar theme at Aberystwyth, 'Inculcating a love of music and fostering directly special talent in the rising generation.' It was a proposal which prompted the two wealthy sisters, Margaret and Gwendoline Davies of Gregynog to provide the finance to establish a Chair of Music and they were also instrumental in enticing Walford Davies to the University College. As he developed his concepts he believed he could encourage, 'A true feeling for the riches of instrumental music and to shift talent away from the narrow competitive ethic into a wider appreciation of good repertoire drawn from the whole treasury of European music.' When those thoughts became plans of action, Pauline and Enid were to play a notable part and Walford Davies encouraged and supported them.

In 1920 they had made Llanarth and its environs their base giving recitals at the University College and joining the orchestra under the direction of Charles Clements, an accomplished organist, pianist and accompanist and a senior lecturer in the department of music. He, in turn, was deeply involved in the planning stages of a huge pageant at the imposing fortress of Harlech castle, which was organised to raise funds for a 'Hall' in memory of the men of Harlech who were killed in the Great War. Walford Davies, one of the musical directors, had gathered a team of writers and musicians, notably A P Graves, the Irish essayist, songwriter and poet, who had played a leading part in the renaissance of Irish literature and music. The event, to be held on August 29th 1920, involved a cast of hundreds including an orchestra, bands, harps, actors, dancers and choirs and called for attention to detail and a military-style precision and organisation.

Charles Clements was detailed to engage instrumental players to form an orchestra, but as correspondence between him and Walford Davies shows, it was a matter of great concern to Charles Clements that local instrumentalists were simply not of the right calibre for 'such a prestigious event'. But Walford was adamant that they did not have the finance to pay high fees for players, and that he, Charles Clements should encourage talented instrumentalists from the University College. Eventually, a fortnight before the event, Pauline Taylor received a request from Charles Clements to join the orchestra and she wrote of her thoughts and determination on the smallest of post cards:

> I am quite willing to play for the pageant especially, as you say, I may borrow a cello from Dr Walford Davies. This will save me a good deal of trouble and anxiety – I suppose I shall find it there when I come for rehearsal. Miss Barber is not here at present but I'm sure she will play too. If the cello is in Harlech and it is fine weather we may be able to save you the expense of a car – by cycling.

The pageant, held over three days in brilliant sunny weather at the end of August 1920, was a huge success. As one journalist reported:

> Daring in its boldness designed to repeat in modern days the town and castle's long and bitter struggles – Royalists, Cromwellians, Parlimentarians, the uprising of Owain Glyndŵr, Queens and Princesses, court ladies and Welsh peasants. Archers, spearmen, men in armour, chain suits, men at arms in gay colours, the castle walls reverberated with the sounds of past events. It began with the scene recalling the lost lands of the legendary Cantre'r Gwaelod (the lowland hundred) and ended with the Coming of Peace.

CHAPTER TWO

A YEAR LATER IN 1921, Sir Walford Davies delivered a lecture to the Cymmrodorion Society hosted by Prime Minister Lloyd George at Downing Street. His title 'Our Mother Tongue, a Musical Policy for Wales' was the perfect platform to elaborate his vision. His opening remarks to the august gathering were, 'Wales is a trilingual country, she speaks Welsh, English and Music and the greatest of these is music.' He went on to elaborate his profound belief that the future lay in extending the range of music teaching in schools and particularly in making and appreciating good melody. His notion of establishing a Welsh Council of Music was financed and accepted, and he became its first director in addition to his professorial role at Aberystwyth. Walford Davies was soon revitalising musical activity in other educational establishments in Wales – schools, communities and the four University Colleges.

At Bangor, he appointed E T Davies, a Dowlais man and a widely respected conductor, teacher and adjudicator, and under his enthusiastic guidance the music department became the liveliest in the University of Wales. Pauline and Enid leapt at the opportunity when The Dorian Trio was selected to work at the Bangor music department to introduce instrumental classics for young people, setting up a series of concert performances and teaching classes. Their appointment gave them a measure of security and a base for building a reputation not only as performers, but also as teachers and lecturers. E T Davies was fully supportive and slowly, the carefully selected programmes they offered north Wales' schools and communities began to appeal, and interest in instrumental music grew. These were happy and fruitful years, but they were not without their difficulties.

There were those teachers who dreaded the words

'chamber music', but with thought and care, Pauline, Enid and Kathleen planned their programmes to meet every listener. They recalled their experiences with these words:

> The simplest of grown-ups and the youngest of children find themselves caught unawares into enjoying trios by Haydn, Mozart, Beethoven, Schubert and Brahms. We invariably include some Welsh music in every programme – even though it maybe two or three Welsh folk songs, often unaccompanied duets for violin and cello, which will open the door to more difficult works. We have had recent proof that Welsh pieces can awaken the creative imagination.

It would be inaccurate to imply that north Wales was a musical desert because they met enthusiastic audiences and often the result of their visits and concerts would be the formation of an instrumental class or, if there was already a musical class or orchestra in existence, they would give guidance, encouragement and help. It was a musical crusade. The power of music was to win and hold the most rebellious spirits.

At the same time they continued to give concerts nationwide at the most prestigious of venues and to the most receptive of audiences. The programme at the Aeolian Hall, London included works by Mozart, John Ireland, Beethoven, 'Scherzo' by Hubert Davies and 'Two Miniatures on Welsh Airs' by E T Davies. Another recital at the Wigmore Hall, brought this comment from *The Daily Telegraph*:

> The Dorian Trio played most carefully and intelligently. Their fine musicianship, however, was more appreciated where care and intelligence found better scope – in the trios of Beethoven and Schumann, which made up the rest of the programme. In these the excellence of the ensemble told with fine effect.

Pauline was a commanding presence playing the cello – red-cheeked, hair cut in a bob of the fashionable style of the 1920s, eyes closed, her body swaying gently to the different rhythms, her face expressing and emphasising her intense involvement and pleasure in the music. Enid was different. Shorter than Pauline, a calm, tranquil presence at the piano, she 'lived the music', with surety and an inner contentment, a quiet understanding smile as she listened to the occasional eccentric image Pauline created during her explanatory introductions. School children who had never seen a stringed instrument such as the violin would listen in wide-eyed wonder at the stories about the violin and the cello and to the sounds they made and the tunes they played. Occasionally at a venue there would not be a piano or maybe there would be one suffering from 'acute old age' and so, they would take a viola to the classroom. Pauline would explain the difference between them, 'mother fiddle,' 'little fiddle' and 'father fiddle,' joining together to play music.

Pauline's nephew, Professor Anthony Bradshaw ventured an opinion about their relationship, the chemistry between them, whether, as very different characters they argued or disagreed. 'They always seemed to get on. Enid would say to Pauline, well, if that's what you want to do, then we'll have to do it. But you always had the feeling she was implying that she reserved the right to withdraw.'

When she stood on a platform delivering her introductions, Pauline appeared to be the dominant character, but she relied heavily on Enid's depth of musical knowledge to draw up attractive programmes which young people would find appealing and informative. In their quest to be entertaining, they would contradict each other, but they both appreciated their individual strengths and weaknesses and could effectively convey their knowledge and joy in music to audiences of young and old. As Enid's niece, Jane recalled, 'Enid was my godmother, a generous and a sympathetic encourager.'

They were to remain at Bangor for almost ten years and as Pauline said when they left in June 1931, 'Ten years is a large slice of one's life, and we have made many, many friends.' But their departure was tinged with unpleasantness. Without giving them prior warning, the new departmental director did not renew their contract. The news came like a bolt from the blue and, although secret staffing discussions had been taking place for a year, they never received a formal explanation. Eluned Leyshon, also a member of the music department and who had been the trio's regular violinist for a year, was so appalled at the decision she expressed her concern and distress in a letter which she distributed to every member of the University College Council. She wrote in May 1931:

> I can bear testimony to the fact that until recently they were unaware of your decision of last year (September 1930), not to renew their appointments. I beg with most sincere earnestness, that the Council should institute an enquiry into the work of the Department, during this and previous sessions. I am firmly convinced that it would be found that both Miss Lewis and Miss Taylor have executed their duties loyally and with devoted skill and that the allegations against them are utterly baseless.

There is no hint of the nature of these allegations. The letter implies a hidden agenda, but it was more likely to be a goodly dose of the well-known Welsh musical disease called *cythraul canu* which can attack without reason. In this case, the cure was a departmental clear out, seemingly due to incompatibility and the director's decision to change policy and direction.

A farewell concert held at St Mary's Church hall in Bangor on June 11th, 1931 was a sell out and attended by colleagues, friends and students. The local paper reported:

> They played Schubert's 'Trio in B flat', Haydn's 'Trio in G', 'The Minuet' from Handel's *Berenice*, Boccherini's 'String Quintet in C' – the trio augmented to five by two local violinists – and finally, the beautiful 'Chorale' from Bach's *Sleepers Awake* which the students in the audience sang with feeling.

Music had been a balm to compensate for the unhappy experience, but Bangor had, without doubt, provided them with many happy memories and soon other doors were opening. They returned to Cardiganshire to discuss opportunities with Walford Davies. He had been an inspirational mentor and, although he had resigned as Professor of Music at Aberystwyth in 1926 to take up the position of organist at St George's Chapel, Windsor, he remained active in Wales as Chairman of the National Council of Music. He told them that there was an opportunity to continue their crusading work in schools and communities in south Wales. They based themselves in a flat at Brecon but they would return to Llanarth when they were within reach of Pembrokeshire and Cardiganshire. Walford Davies kept in close contact with them and they, in turn, would regularly report their activities to the Council. Enid failed to attend one meeting due to her mother's illness and within days she received this letter from him:

> My Dear Enid,
>
> We missed you and your cheery and vigorous account of Trio doings at the Council, but your splendid work of the session was fully evident and you can imagine how it delighted my heart.

He goes on to ask for her help is assessing the qualities and needs of two students who have written to him. One is looking for financial assistance and Walford Davies had sent him five pounds; the other, a girl, is looking for a teaching post:

> Peggy has a fine voice, but she is not good enough for a secondary school teaching post. She ought to be able to teach in an elementary school and we ought, somehow, to find her a niche but it is one of those causes that are desperately difficult to help – no initiative, a beautiful voice and a certain amount of ambition. I am awfully sorry to trouble you, but I know you will be the link and let me know in complete confidence what you think is the best way of helping them to help themselves.
>
> Yours ever affectionately,
>
> Walford Davies

Their experience and judgement as teachers and professional players was much in demand and they frequently engaged the violinist Violet Palmer when they played chamber music concerts at well-known concert halls and suitable venues. 'The Dorian Trio deserves to be specially commended for its enterprise for including the work of Gabriel Fauré, the trio 'OP.120' at the Wigmore Hall. One felt grateful for the sensitive and thoughtful performance.' So wrote *The Daily Telegraph* in 1936. Another music critic of *The Sunday Times* reported, 'The trio played with refinement and consideration for one another and the music.'

During this period in the early 1930s they travelled extensively in the five south-west Wales counties to fulfil their obligation to schools and communities. The industrial communities were very different from the rural areas of north Wales, although they noted painful similarities. They wrote of their experiences in the periodical *Y Cerddor* of the 'desolation of large industrial districts and of the countryside,' as the economic downturn took effect:

> In different senses they are both distressed areas due to high unemployment and the economic slump. The distress in the one case comes from overcrowding and the conditions of town life which hamper the free growth of body and character, and in the other, from the hopelessness of an ignored minority struggling against the neglect of agriculture and lack of opportunities.

They travelled to Llanelli, Swansea, Neath valley and Brynmawr, and later they recorded their impressions of a visit to the unemployed at the Mens' Club at Dowlais where the distress was intense:

> Many of these men and boys have never earned a day's wages in their lives. No wonder that there is moral, mental and physical disease among them. On a bitterly cold night it made a dramatic scene – the large, bare, grimy warehouse-loft which does duty as a club-room, and those rows of eager, yet haunted, hungry and

desperate faces listening and watching intently as we played. From the first note we had perfect silence as we played and explained the music as we went along. Faces grim and strained became for the time being gentle and relaxed.

They formed a music class there and returned every term to give encouragement:

Country children are reserved and shy. They scarcely know whether to applaud or not. They show their appreciation by their positive stillness. Unconsciously they absorb all the blessings of life as they walk along lanes and fields to school every day; they are in close touch with this 'mysterious universe'.

At times they felt their work was becoming a social critique:

In the remoter parts of the country there is just a chance before it is too late of laying the foundation of a taste for what is true and strong. It is a desperate race against time, for the film talkies, jazz, the wireless and gramophone indiscriminately used are rapidly destroying the countryman's capacity to respond to what is of lasting worth in life and art. This labour cannot be in vain, for we are preparing for the future.

This philanthropic dedication to their educational role brought appreciative comments from prominent educationists in south Wales and beyond. F A Cavenagh, Professor of Education at King's College London complimented Pauline Taylor's explanatory introductions, 'Most helpful for understanding both the composer's place in musical development and the features of each composition. I can cordially recommend The Dorian Trio to any school wishing to develop their students' love and understanding of music.'

W Emrys Evans, Director of Education for Breconshire wrote, 'The educational value of these visits is beyond question,' and H J Lewis, Director for Cardiganshire commentated in similar vein adding, 'I trust that many more schools in our county will, in the near future, have the privilege of hearing this excellent trio.'

These lecture concerts became an established feature in the school year, and generations of children in south Wales in the 1930s, 40s and 50s recall those concerts which occurred at least twice in a school year. A few would be enthused, the majority listened politely – 'because we were told to' – and for them these musical interludes tended to be sessions to be endured rather than enjoyed and appreciated. At Ardwyn Grammar School, Aberystwyth, my alma mater, The Dorian Trio's appearance caused interest and amusement in equal measure because of Pauline Taylor, tall, red-faced and grey-haired, with her English accent adding a touch of eccentricity to her stories and comments about composers and compositions. Her clothes conveyed a colourful and to us a bohemian lifestyle in complete contrast to Enid Lewis, smart, well-groomed and structured in textured tones at the piano. The cello, a beautiful instrument occasionally, caused Pauline a few problems – her skirts were not quite long enough to prevent rows of oggling boys the opportunity to count the number of times they caught a glimpse of her shiny bloomers. As a school friend of mine once recalled of those occasions, 'I saw things I should not have seen.' And he left his musical memories at that.

Pauline and Enid were an integral and important part of the musical life in Wales in the 1930s making the parish of Llanarth their base and home. The village is one of the older settlements in Wales, on the crossroads where the A487 road from Cardigan to Aberystwyth meets the B4342 which leads to the picturesque seaside village of New Quay. Ship building and seafaring has played a prominent part in the lives of most families; in the first twenty years of the nineteenth century, 31 ships were built in New Quay, and most farmers worked part time as ship builders or as fishermen during the herring season. But Llanarth's main claim to fame took place four centuries earlier when Henry VII encamped his forces on his second night after landing at Milford Haven on his way to do battle with Richard III at Bosworth. Henry stayed at Plas y Wern, a small mansion in Llanarth and was entertained hospitably by Einon ap Dafydd Llwyd and a plaque in the west wing, the oldest part of the house, denotes that event.

The land rises gently to about 600 feet above Cardigan Bay at Blaenwern – a large solid house which stands in its own grounds. A lane bisects the farm buildings from the house; it's one of those lanes that winds without a signpost, through undulating bare hills and wooded dingles and eventually leads to the close community of the large village of Talgarreg. This was the area, known as Banc Siôn Cwilt, that Enid had known since childhood, where her mother had grown up and where she had spent many school holidays. Two farms, Blaenwern and Gofynnach Fawr, formed the heart of an estate which Enid's grandfather had built up over the years buying smallholdings in the area until, by the end of the century, he had acquired 500 acres. When he died he was deeply in debt, his business ventures acquiring Cardiganshire farmland did not match his purse, and many of the mortgages were surrendered. But Enid's grandmother, well into her nineties by then, remained at Blaenwern to continue farming with Johnny, one of Sadie's siblings. It was here that Enid would stay on her return to Llanarth. In fairness, land values had fallen but Enid's father, John Lewis, whose parents hailed from Cardiganshire and always the astute businessman, bought three of the smaller farms, Penlon, Penrhiw and Rhydfawr as an investment for his wife.

Two years later in 1933, Enid suffered two tragic losses. Her father died after a long illness in the first week of January that year, and in September, her mother, the indomitable Sadie, who had also not been in the best of health, also died. On her death, Enid inherited the three holdings with about 100 acres of land and, although there were sitting tenants in these farms, she felt she was gifted to remain in Cardiganshire and to put down her own roots in the Welsh countryside.

Another member of the Thomas family, Sadie's brother, was to influence them at this time. He was the respected and colourful pastor, the Reverend Joseph Lloyd Thomas who had recently retired from the Unitarian ministry at the Old Meeting House in Birmingham. He had been a reforming and inspirational minister and, at one time, his influence went far beyond the Unitarian church with his aim of combining the intellectual freedom of the 'free church' tradition with Catholic worship. He called himself a free Catholic priest and in 1919, within the Unitarian ministry, he founded the Society of Free Catholics. Ten years later, he edited the Free Church Book of Common Prayer to include the Nicene Creed, canticles and psalms. He may have been following a personal quest, creating intellectual mayhem as he did so, but he was not without his supporters and admirers. There were many who were pursuing similar objectives but, in truth, the new movement languished in its infancy. In 1932, Joseph Thomas and his family retired from the ministry to a smallholding, Y Bwthyn in Llanarth, and although his dream of unifying the denominations did not materialise, he was soon involved in Cardiganshire community affairs. During the next fifteen years he was elected a County Councillor, made Chairman of the highways committee, became one of the founders of the Ceredigion travelling library and served on the county's education committee.

When Enid's grandmother died in 1934, Uncle Joe became an even greater influence on family affairs – dealing with the estate, the will, debts, and negotiating an agreement between his five brothers and sisters. Sadly, the two remaining farms of the estate, Blaenwern, the family home, and Gofynnach Fawr were sold, the moneys divided between the six remaining sons and daughters, and Enid, Sadie's daughter, inherited her share. Enid kept the three smallholdings directly bequeathed to her by her mother, and one of them, Rhydfawr, became her Cardiganshire home for the next twenty-five years.

Uncle Joe was delighted to meet up again with Enid who, through his Unitarian ministry, had been influential in assisting her, through his contacts in Budapest, to arrange her two years studying in Hungary. Enid found accommodation with other relatives in Llanarth until Rhydfawr became available. Uncle Joe offered Pauline a room at Y Bwthyn and in many respects, the next four years were times when she crystallised her thoughts and ambitions through conversations with her surrogate uncle. He was the father she had missed so much during her formative years and he was also a great comfort to her when her mother, Florence, died in 1934. Pauline was at last putting down roots in Cardiganshire. Uncle Joe was a man of letters with a deep knowledge of Welsh history, heritage and culture. He was an inspiring teacher and he instilled in Pauline an enduring affection for the land and the people of Wales. He had never been afraid of espousing the minority view and supporting unpopular movements. He ventured to express publicly his sympathy with the enemy during the Boer War and he had supported the suffragette movement by chairing fiery meetings for Christabel Pankhurst. Pauline grew to admire his conviction, and learned to debate her views without rancour. When Uncle Joe was elected a County Councillor, she accompanied him on horseback to his committees and meetings and he, in turn, would give her and Enid every encouragement in their musical education work. They even talked of setting up a pony trekking business and when Pauline made her first investment (she bought a Welsh cob mare), she stabled it in Uncle Joe's stables. At last, she felt a sense of permanence coupled with a strong possibility that her childhood dream would one day become a reality.

CHAPTER THREE

IT WAS A CHANCE meeting at a concert in Cilgerran that brought Barbara, Pauline and Enid together and their developing friendship was to become pivotal to their future ambitions. Barbara Saunders Davies was wealthy, born in 1907 to a life of privilege on the Pentre estate in north Pembrokeshire. Her early life was one of plenty: taught by a governess at home, followed by two years at Malvern Girls' College and a finishing school in Switzerland, travelling, riding, fast cars – she could have personified the lifestyle of rich landowners. But in her twenties Barbara had begun to think of social and academic improvement, using her experiences to read widely and to pursue scientific and cultural knowledge. She studied geology, genetics, astronomy, history, languages and music, especially the piano, and although she never regretted not having a formal education, she followed her own interests and immersed herself in the pursuit of knowledge. When she first met The Dorian Trio in Cilgerran in 1936, she had just returned to Wales from Paris where she had been studying the piano with Nadia Boulanger, the most influential composer, conductor and music teacher of the twentieth century. Attending the village hall concert she expected an amateurish performance, she wrote:

> I was pleasantly surprised to hear a very musical and professional programme, which I thoroughly enjoyed. Hearing that they regularly gave recitals at secondary schools in south-west Wales, I invited them to stay at my home, Pentre, Boncath, when they were in the area. So began a long collaboration.

Barbara was brought up on Victorian customs and values. She was not indulged or spoilt and her enquiring and retentive mind powered her search for knowledge with a scholarly approach to reading and travelling. In that sense she was very much like her father, Arthur Picton Saunders Davies, who preferred to study and travel than concern himself with the affairs of land ownership in west Wales. His main interest was studying different religions, and he travelled to all parts of the world in his search for a deeper understanding. It was during one of these journeys to the USA that he met Mabel Woodruff, a teacher of Christian Science in Boston. There followed a whirlwind romance and, as Arthur left for home, he vowed to return within weeks to marry her. Mabel, hopelessly in love with a man at least ten years her senior, swiftly made the wedding arrangements and an announcement in *The New York Times* proclaimed on the 12th of September 1899, the marriage of Arthur Picton Saunders Davies and Mabel Daughaday Woodruff of Seattle, teacher of Christian Science:

> The Christian Science ceremony was performed at the home of the Mayor B D Webber. After the wedding, Mr and Mrs Saunders Davies left for New York whence they will sail for England tomorrow. They first met two months ago. Mr Saunders Davies is the proprietor of several landed estates in Wales.

They arrived at Pentre to find the family not best pleased with the union, but that attitude had little effect on them as they began their married life in the new century. Their first child, a son, Arthur Owen was born in 1901 and Barbara, his sister in 1907. The house and the extensive gardens in the wooded valley, a mile or so from the river Teifi, had been greatly enlarged and stone-cased by Arthur's mother, who added an attic storey and chapel in memory of her husband. It was lavishly furnished with glass chandeliers, large ancestral paintings and suits of armour strategically placed on each landing. The walled garden and sweeping lawns added to its charm and tranquillity. The house remains standing and occupied today, but according to historian Thomas Lloyd, who wrote in his publication, *The Lost Houses of Wales*, 'The

33

chapel, a fine period piece, with painted walls, heraldic floor tiles and stained glass, but savagely bulldozed in 1950 without salvage, together with the whole front pile of the house.'

I went in search of the mansion where Barbara grew up and the home which meant so much to her. Luckily it was a sunny spring day and I approached the village of Boncath (the postal address for Pentre mansion), which is a mile or so to the south of the river Teifi bordering the counties of Pembrokeshire and Ceredigion. I stopped at the local shop and the manager was able to direct me through the intricate web of high hedged lanes – the countryside blooming in the morning sun with wild flowers: bluebells, foxgloves and red campions mingling with new growth of a lush variety of hedgerow grasses. At a gateway I stopped to look to the right, and through the trees I could make out a number of high, grey-stone farm buildings almost hidden in the wooded valley. At the next lane, I turned to drive towards a set of large iron gates. I had arrived at Pentre. In front of me, a driveway led ever deeper into the wooded valley and on the left, a sign on the gate of a small dwelling said 'Pentre Garden Lodge' and on the right, a newly erected, garish, jaunty sign in bright red and yellow paint pointed to, 'Pentre farmhouse and cottages'. I learned later that these were for sale.

The imposing iron gates were wide open, the driveway flanked with thick rhododendron bushes in full bloom, but spoilt by a large sign which read KEEP OUT, PRIVATE PROPERTY. I drove slowly down and as the drive veered to the right, I saw another sign nailed to the branch of a tree, with a more sinister warning – BEWARE, CCTV CAMERAS IN OPERATION DAY AND NIGHT. No doubt, I was being watched. I drove on slowly to find out why, and around another gentle bend, the drive opened out to reveal an unimposing large grey building. This was Pentre mansion and Thomas Lloyd's words 'savagely bulldozed' underlined the fact that the building lacked any charm or a distinctive feature. Parked in front of it were four large black cars, two Audis and two BMWs; there were reels and reels of steel piping and cables strewn along the gravelled drive and, on the south-east side, an enormous satellite dish almost as high as the house. This was an unexpected, slightly sinister scene in this tranquil wooded Pembrokeshire valley. A deep silence descended on the scene, voles and moles stopped digging, rabbits and hares held their breath wide-eyed like cameras watching my every move. I looked around with a growing feeling that I would soon be bundled into a Tardis and thrust by rocket into another world. With a sense of irrational foreboding, I turned the car to face the drive in case I needed a quick get-away, locked it and walked to the front door. The grounds were well kept, lawns mown and borders clean, but the house looked bereft, the sun highlighting its stark greyness and lack of charm, windows tightly shut and darkly shuttered. Even the birds seemed to have momentarily lost their voice, pausing in the silence.

I rang the door bell. It was not the sound or the tone of a welcoming bell but a high pitched piercing whistle, a siren, causing the birds to cry and flutter. I waited while they settled to enjoy the sunshine. I looked down and a marble memorial plaque leaning drunkenly on its side against the front wall caught my eye, the words as I remember, 'Erected in memory of Arthur Henry Saunders Davies by his loving wife.' Before I could memorize the dates, the door opened, no more than two inches and then wider to reveal a dark-skinned Asian lady in a long dark skirt and a gillet over a sweater. She pushed her long hair away from her face, and then pulled the door almost to a close behind her. 'Yes?' her lips scarcely moved in an inscrutable face. 'I'm not a trespasser,' I said lamely. 'I've come to see the home of the Saunders Davies family who lived here for two hundred years until the 1950s. I'm doing some research for a book.' She was unimpressed, she didn't blink. We looked at each other.

'Yes,' she said curtly, 'I don't know anyone of that name,' her dark brown eyes hard and unfeeling. 'Well you've got a memorial stone outside your front door.' Realising that I was on a helpless and hopeless mission, I stumbled on, filling the silence with suitable inanities, 'I wanted to see the house, to see where Barbara Saunders Davies was brought up. Who lives here now?' At that point a telephone rang, she half turned to the house, 'I have to answer, excuse me.'

She disappeared and closed the door as noiselessly as she'd opened it. I stood waiting, staring at the door, not daring to move for two minutes or more until the door opened again, wider this time and her tone of voice when she spoke was more aggressive, 'Sorry, I cannot tell you anything. No-one knows.' And suddenly I knew she'd been given orders, such as, 'Get rid of her.' And she more or less summoned me to leave, 'You saw the notices.' I nodded, 'Yes.' She stood her ground and for half a minute, so did I. I replied with all the authority I could muster, 'I'll go. Sorry to have disturbed you, I've nothing to hide.' And I blurted again, as if to emphasise my innocence, 'I only wanted to see an important family home in Pembrokeshire, the house where the Saunders Davies' family lived for two centuries.' And as I turned to walk to the car, I looked at her, the door now wide open. She stood 'full frontal' as if she was taking in the air and the sun. I couldn't help myself, so I said, 'I hope you enjoy your secret life.'

She didn't blink or smile. She stood motionless watching me as I drove up the drive out of sight, the cameras I have no doubt following my every move. At the gates I turned left obeying the 'Cottages for Sale' sign to follow the narrow lane towards the farm buildings, passing a walled garden on my left. There was no one about and oddly, no warning signs. I stopped and walked around in deep silence until a tractor in the far distance began to cut a field of grass for silage. Whereas the house had looked grey, bereft and unkempt, the barns, stables and other grey-stone buildings stood high and proud although obviously unused, beautiful in their lonely solid splendour, a reminder of life on large estates and the golden age of solid stone buildings, farmworkers, gardeners, horses, phaetons, landaus and coaches, housemaids, butlers and cooks. These were the buildings 'For Sale'. Generations ago they had been a hub of activity, but today, when I enquired in the villages around whether anyone remembered the Saunders Davies family and the Pentre estate, no one could answer.

I had lunch of Teifi sewin at a homely restauraunt in Cilgerran. 'Do you know anything of Pentre Mansion?' I enquired of the lady owner, and after a long pause she replied, 'No, no, I'm sorry, I don't. Where is it?' 'A mile up the road.' I failed to find anyone who could tell me anything of the present owners or about the goings-on in the house, not even an elderly lady who told me she had lived in the area 'all her life' as she squinted at me in the afternoon sun as I mentioned the name Saunders Davies, 'I don't recollect anyone of that name.' I turned for home and I had a sneaking feeling that they, the mysterious owners of Pentre and their huge satellite dish, knew everything about us. The occupants were Skycom Telecom Ltd. Part of the satellite information system and World Meteorological Organisation, but the secrecy and the attitude of Messers J Shah, I Maclean and P Wynter makes me think there's more to it than looking at the clouds and forecasting the weather.

*   *   *   *

Barbara's father died in 1922 and his wife Mabel remained at Pentre until 1947. In its heyday there were at least forty rooms, a large library, a chapel and a music room and when Barbara turned eighteen (three years after her father's death), she had a glorious 'coming out ball' following the tradition when young girls, the debutants of the upper classes, were introduced to society and were eligible to marry. The estate included land of some 10,000 acres in three Welsh counties, Pembrokeshire, Carmarthenshire and Cardiganshire, with an annual rental of almost £8,000 in the mid-nineteenth century. Barbara's brother, Arthur Owen, being six years older, inherited the estate when their father died, but he turned his back on land management to become a motor racing driver and the highlight of his career happened in 1931 when he came third in the Le Mans race. When Pauline and Enid established their friendship with Barbara, they soon realised that, like her brother, she loved speed and the narrow Pembrokeshire lanes echoed to her highly tuned motor car being driven to its limit. Her great sport was racing the small country train called *Y Cardi Bach* which chugged along the railway line adjoining the western side of the Pentre estate between Whitland and Cardigan. From everything I've learned of Barbara's driving, it was not a contest.

The friendship between Barbara, Pauline and Enid was kindled through their interest in music, and forged by their academic backgound and their pursuit of knowledge. They were often invited to stay at Pentre and occasionally Barbara would visit Llanarth to ride with Pauline and Uncle Joe. She wrote of her memories of their music-making:

> Just before the war Pauline and I discovered recorders and thought they were ideal instruments for schools to use to teach the reading of musical notation. In those days music was taught from reading Tonic Sol Fah, so we introduced recorder music into their programme. I took up the oboe but never the violin. It was unfamiliar, such a small sound compared with the familiar choral singing and works like the Messiah, but half a dozen regular concerts may have sown seeds for later appreciation. One little girl went home and told her parents with some excitement, 'Mummy we've seen the wireless.' Another appreciative comment after a concert came from a small child, 'Oh, I love that piece, the Swan. There is Miss Taylor with the cello floating on the water and Miss Lewis on the piano, she is the swan's legs.'

If this appeared to be a rather 'elderly' picture in the mind, it was probably due to Pauline's hair which had turned white in her early twenties.

Barbara lived with her mother at Pentre and on her father's death she inherited £30,000 annually from the estate in addition to the silver, china and glass. She was tall and elegant, always wearing classic, well-cut clothes. She was not a follower of fashion but she seemed extremely glamorous and although she loved Pentre, she appeared as an outsider and 'very English' in the Welsh-speaking communities of north Pembrokeshire. The clothes hid a rather shy, withdrawn personality but when she spoke she had a beautifully modulated voice and a gift for language. She may have appeared diffident but there was an inner strength and determination about her. If she set her mind to do anything, especially if it entailed research, she approached it with an organised, studious mind. She enjoyed Pauline and Enid's company because of their enthusiasm, love and knowledge of music and their enjoyment of the countryside.

The Dorian Trio continued to give concerts until the outbreak of war but Enid was already gaining a reputation as an effective teacher of pupils studying the piano. She had a small income from her legacies, but Pauline had no other source of money other than her shared earnings from concert and lecturing fees. She and Barbara had forged an alliance because of their interest in horses, in particular the Welsh pony and cob breeds and other Welsh breeds. Enid did not have a similar interest but there was a strong long-standing friendship and bond with Pauline and, although the outbreak of war curtailed their activities, they did not give up their work as The Dorian Trio.

In January 1940 another opportunity presented itself to Pauline. The Council for the Encouragement of Music and Arts (CEMA) was established, the first government-funded body dedicated to promote and maintain British culture. It was an initiative regarded as a milestone in artistic history, with a mission to maintain the highest musical standards while keeping them accessible to the general public during wartime, and to stimulate amateur music-making. The self-chosen motto was 'The best for the most'.

Pauline was appointed as the organiser for south and west Wales, and in many respects it continued the educational work that she had been doing with the Trio. But there was an added motivation. She was to gather around her a team of 'music travellers', as they came to be called, to arrange concerts by professional musicians and to foster music-making, all designed to sustain people's morale during a time of war when it was most needed. The majority of 'travellers' were women, as were many of the professional musicians and entertainers. There were memorable daily lunchtime recitals by the pianist Myra Hess at the National Gallery in London, readings by Dame Sybil Thorndyke at locations during a tour of the mining valleys of south Wales, eminent singers such as soprano Isobel Baillie giving recitals and joining chapel choirs in performances of works from the oratorio repertoire. Performances that were often portrayed as 'soothing the nation at war,' or, as one historian recalled, 'When women were portrayed less as workers but more as symbols of the Britain that British men fought to protect.'

Pauline was in her element, with her team cajoling reluctant musicians to come 'all the way' to Wales, and arranging concerts and events at venues, large and small, in all parts of south Wales. It was a mission which brought to mind the concerts she and Enid had given in the Midlands during the Great War, 1914–18. The Trio played their part in the CEMA scheme; their flat in Brecon was an ideal location for travelling, often in difficult circumstances through the blackout, without street lighting and without many road signs in rural areas. Artists new to the area got hopelessly lost, using far too many petrol coupons from their precious allocation. An added impetus to the early concept and founder members of CEMA was given by three eminent Welshmen, Dr Thomas Jones, civil servant, Cabinet Secretary to four Prime Ministers including Lloyd George – a man of a thousand secrets; William Emrys Williams who was Director of the British Institute of Adult Education and Sir Walford Davies, 'an inspiring loveable optimist' as someone once called him. They were keen to encourage amateur music-making, whereas Sir Kenneth Clark, another founder member, felt strongly that the idea of education was simply to offer artistic performances of the highest standard.

There were often breaks in the schedule which allowed them time to return to Llanarth and to visit Barbara at Pentre. Life had changed on the estate. The military authorities had commandeered the mansion as an Auxiliary Hospital and Convalescence Home for sick and wounded servicemen. They occupied one half of the houses and the family lived in the other half. A snapshot of life on the estate at the time was recorded by Helen Ogilvie from Barrow-in-Furness who joined the Women's Land Army in 1941 at twenty-two years of age and who was posted to Pembrokeshire. She was employed in the horticultural section at Pentre:

My work was mainly in the greenhouses, thinning grapes on the vines, pollinating the peaches and nectarines with a rabbit's foot and making sure everything was well watered. I also picked the soft fruit, climbed the apple trees in the orchard, packing the surplus fruit to take to the shops in Cardigan. The staff had all been retained – the cook, the maids, Lloyd the chauffer who would convey the servicemen to and from the local railway station in the family limousine. The service personnel were cared for by Voluntary Aid Detachment nurses under the command of Matron, with an Army Medical Officer attending weekly. Sunday morning services were held in the chapel attached to the mansion with one of the nurses playing the organ and the local vicar conducting the service. Miss Barbara Saunders Davies, the daughter of the family and about ten years older than me, bred beautiful palomino horses. These were a lovely golden colour with cream mane and tail, and after work I spent many pleasant hours being taught horse riding by her.

Helen Ogilvie spent seven years at Pentre and in 1945, a month after Armistice Day in Europe, Helen met her future husband. Royal Marine Stanley Ogilvie came to Pentre as a patient to recuperate from war wounds and attacks of malaria. But it was two years later, in January 1947 that Helen was released from the Land Army and left Pentre with mixed feelings. They married in August that year; identical twin daughters, Dilys and Glenys were born two years later and when Stanley was offered a post in the civil service at Swansea, they returned to Wales.

CHAPTER FOUR

PAULINE AND BARBARA BEGAN discussing plans for a joint farming venture based in the Llanarth area. Pauline, full of her customary dynamic energy, was the driving force. She was already the owner of two Welsh cob mares and following her labours with the CEMA, she returned to Cardiganshire to ride and discuss with Barbara the possibility of a partnership. Barbara was keen to develop a Welsh cob stud, her considerable scientific knowledge of inheritance and the genetics of breeding enabled her to identify the characteristics needed to breed cobs of quality and to develop different bloodlines and types. Pauline had a rather different approach. Fired with enthusiasm from her discussions with Uncle Joe, she was consumed with a mission to promote the potential of all Welsh native breeds, an unfashionable idea in the late 1940s – Welsh Black cattle, Llanwenog sheep, Welsh pigs, Welsh dogs and, of course, Welsh cobs. She had already mated her pure bred mare called Firefly to the famous stallion of Cardiganshire, Brenin Gwalia – the king of stallions. This was the beginning of an odyssey into the closed world of Cardiganshire cob breeders, up until then very much a masculine preserve. They were smallholding farmers, subsistence-living on relatively poor land, and the bonus of possessing a Welsh cob champion stallion added to their income during the equine mating season between April and July.

It was a time of dramatic change in farming. The old way of life, so dependent on horse power, was rapidly changing. The development of the internal combustion engine from the end of the First World War had left little doubt that the horse had had its day. The rate of change from horse to motorised transport in towns and urban areas was dramatic but investment in new machinery in rural areas was protracted, especially on small family farms, and when the first Ferguson tractor came off the production line in July 1946, it revolutionised agriculture. It was affordable, strong, manoeuvrable and flexible with the added advantage that all the implements were locked and controlled by the tractor's own hydraulic power system. The pace of life had quickened.

The next step was to buy land. Pauline had no funds but Barbara was prepared to invest and the first purchase was Dolau, a typical Welsh smallholding of 50 acres and later another, Frongoch – similar in size, in the area overlooking the sea near New Quay. Both transactions amounted to £850 to acquire 100 acres of land. These small enterprises run by tenant farmers enabled them to begin developing their pedigree livestock programme. Neither Pauline nor Barbara had much experience of practical farming, but their tenants were countrymen and for the first two or three years, until the end of the war, they continued to farm and to look after the cattle and cobs. Apart from the two mares, their next purchases were ten Welsh Black cows which were placed with the cobs at Dolau, keeping Frongoch available to rear young stock. Although their minds were set on breeding pedigree lines, they both realised they needed to be practical and commercial about their farming enterprise. The agricultural industry was showing signs of recovery following the depression and the crippling slump in land values and food prices of the late 1920s and 30s. Farmers at last had secure markets for their products when, in the 1930s, the government established marketing boards for milk, meat and arable crops.

Enid Lewis had little or no interest in farming and she realised that Pauline, as her work with the CEMA came to an end, would be spending more and more time at Llanarth. In 1945 Enid decided to concentrate on teaching and she took up an appointment at the Guildhall School of Music in London. There was no animosity – she retained her friendship with Pauline and Barbara, and she kept her cottage, Rhydfawr. It was by no means a total break from the past; The Dorian Trio would remain an integral part of their lives, continuing to give concerts in schools and communities in Wales and England for the next ten years. This was to be Pauline's only regular source of income, and a well-kept diary for the 1950s documents how busy they were – on occasions visiting three schools a day – with notes of programmes they played at each location. Their schedule on November 8th, 1950 began at 09.45 at Henllan Modern School in south Cardiganshire where they played: 'Plas Gogerddan'; Violin and Cello: 'Bourrées 1 & 11', Bach; 'Trio in G', Mozart; Violin: 'Air on G String', Bach; Cello: 'Minuet', Boccherini; 'Trio', 'Suo Gan', 'Nos Galan'. They travelled to Llandysul Grammar School by 11.30 where they played: 'Trymder'; Violin and Cello: 'Bourrées 1 & 11', Bach; 'Trio in G', 'OP 1', Beethoven; Cello: 'Elegy', Fauré; Violin: 'Gavotte', Bach / Kreisler; Trio 'Nos Galan', 'Morfa Rhuddlan'. At 2.30 pm they were playing at Lampeter with a similar programme except for a piano solo, Balfour Gardiner's 'Noel'. The following day, November 9th at 09.45 they were at Aberystwyth Modern School; at 11.15 it was Ardwyn Grammar School, and at 2.45pm they had travelled 40 miles to play in Newtown. Their music schedule was strenuous but satisfying. Enid was able to arrange her teaching schedule at the Guildhall to accommodate the Trio tours but Pauline, although she enjoyed the concerts, longed to return to her animals and the peace of the Llanarth countryside. Firefly, her first Welsh cob mare was registered in the Welsh Pony and Cob Stud Book and for the first time she used the prefix Llanarth to identify the stud. The Stud Book is the breeders' bible, as it records the breeding details of all sections of the Welsh Cob and Pony breed.

'Llanarth Firefly' had been mated with Brenin Gwalia, a revered premium stallion, owned by David Rees, one of the few Cardiganshire breeders who had struggled on his smallholding to keep the premium cob breed thriving. Wynne Davies, the breed's expert historian and practitioner has chronicled the critical years in the history and development of the Welsh cob:

> Cob stallion numbers had depleted alarmingly during the war, a few experts believe almost to the point of extinction with only about five pedigree stallions remaining. Cardiganshire was the home of the Welsh Cob, and David Rees was the third or fourth generation of a family whose lives centred on the native Welsh breed. They never parted with their stallions; they were always buried on the smallholdings. Another breeder, John Evans, lived together 'crofter fashion' with his champion stallion, Cardi Llwyd, on the edge of rough moorland listening to his every movement. The intense rivalry that existed among the owners of premium stallions was palpable and their followers guarded their secrets.

At the turn of the century there were many thousands of working cobs in Wales which had been bred to sell to the Army, for the milk delivery, breweries and newspaper trades in towns and cities. It was a lucrative business, but that source of income disappeared with the growth of motorised power. It became obvious to Barbara and Pauline that to survive in a changing world the Welsh cob needed a new role if it was to survive and prosper.

The area often designated as the country home of the Cardiganshire Welsh cob is relatively small. A triangular tract of hinterland land between Aberystwyth and Aberaeron forming a kind of plateau, 300 to 600 feet above the more productive coastal belt of Cardigan bay, a land of hills and dales which had been reclaimed, drained and enclosed by the labours of smallholders in the mid-nineteenth century. It is the heart of the uphill struggle by farmers to obtain a living from the land – a land naturally devoid of lime and phosphate. The surpluses of this store stock area of Cardiganshire were exported, store cattle and sheep, weaned pigs, breeding ewes and working horses, both light and heavy but especially the Welsh cob. There were between twenty and thirty studs, the majority keeping just one stallion to enhance the farm's income. Owning a good premium stallion gave the farmer status and as one well-known breeder, Ifor Lloyd remembers, 'The stallion was a symbol that the farmer was king of his own domain.'

Barbara and Pauline were entering a keen, competitive world but Pauline had been well taught by Uncle Joe. She realised that to be accepted as serious breeders, they would have to challenge the status quo – that their kind of equality and achievement was only possible by meeting prejudice and scorn with knowledge and courtesy. To succeed in this closed secretive world it was imperative to seek advice and to please two well-known experienced owners. With the bit clenched firmly between her teeth, Pauline followed her first visit to David Rees to mate Firefly with Brenin Gwalia with a visit to his rival, John Evans, at his remote farm. This time she wanted to buy another mare. John Evans may have been curious, he could have been cautious, but he could only admire Pauline's serious intent and determination, and at the end of a day of hard bargaining, she had clinched a deal. She had bought another mare which she named Llanarth Kilda. From that day, they were welcomed and helped by both gentlemen. They had recognised ladies of quality and of serious intent and as they told another breeder, 'They knew a thing or two – those ladies.' Within a few months, a filly foal was born to Firefly; they named and registered her as Llanarth Vega because of the star on her forehead, and this was the first step towards their Welsh cob breeding venture.

As the war ended, another, much bigger farm, Gilfachreda, in the same locality came up for sale and Barbara made another investment so that, at last, they could begin developing a stud and to farm on their own. From the beginning it was clear that the financial input for the business was to come from

Barbara and it was to remain so throughout their partnership of seventeen years. In many respects it was a loose, amicable arrangement without a formal business agreement, based on mutual interests, respect, trust and friendship, and although they argued fiercely, their aspirations for the enterprise were binding.

Their day-to-day working responsibilities depended on the number of livestock breeds they kept. Barbara took over Welsh cob breeding, and Pauline was responsible for the Welsh Black cattle, pigs, sheep and the corgis, and I suspect they shared the tasks associated with arable crops and grassland reseeding. Pauline often said that to make cob breeding a viable proposition in the days when the horse was losing ground to the petrol engine, they planned a mixed agricultural economy which was based on cattle and pigs and, as time went by, the Welsh Blacks gave the stud security. Len Bigley, who was later to work and play a leading role in the development of the stud for many years told me, 'Pauline looked after the cattle and the farm, Barbara dealt with the cobs and paid all the bills.'

'All Welsh breeds have certain characteristics in common,' Pauline once told a journalist, 'Characteristics which are of the utmost value in these austerity post war days with soaring prices for feeding stuff. Production with strict economy must be the rule,' and she went on to qualify their choice with one other fact, 'Our decision to maintain Welsh cobs might have proved disastrous had they not, like our own Welsh Black cattle, been able to a large extent to look after themselves.'

The land around Llanarth was not considered to be the most fruitful – it was classed as marginal land, typical of large tracts of the Welsh landscape between 500 and 1000 feet above sea level, some of it marshy, parts rocky but most of it unable to sustain heavy cropping or grazing without adequate drainage, a heavy input of lime and other fertilisers. There was little shelter; few hedges survived in good order on the open moorland. Hawthorn hedges, battered and blown by the westerly winds from the sea were stunted, bending to the course of the wind, their tops flattened as though they had experienced a manicure session. These were areas where Welsh breeds of livestock thrived because of their adaptability and it is no surprise that the Welsh cob and native Welsh Black cattle adapted to conditions where the south-westerlies and rain regularly blow from the Atlantic. In mediaeval literature, the Welsh cob is described as an animal of:

Fleet of foot, a good jumper, good swimmer and able to carry a substantial weight on its back... The body's conformation had to include a chest with a spread to match the valour of the heart; well-sprung ribs, powerful loins, neat sharp eyes, a dish face tapering past flaring nostrils to a fine muzzle, the whole head expressing intelligence and generosity of temperament. For strong free movement the cob needs tendons of steel, bones of a density seldom found in other breeds, powerful muscles and tough hooves conveying an animal of symmetry and elegance.

Pauline, an accomplished rider thought them to be a delightful mount, sure-footed with free movement at all paces and a born jumper. 'They are full of innate intelligence, quick to learn, reliable, yet eager in spirit.'

Dafydd Edwardes, who lived and worked in the hinterland of Cardiganshire cob land recorded his sentiments in the early 1950s at the passing of the horse and trap days, now that motorcars ruled the road, 'The Welsh cob is the handy man of the equine race; as tough as whipcord and wire, no day too long, he can cope with all sensible work that fits his height and weight.'

One event, which turned out to have a lasting influence on the Llanarth cob breeding programme, occurred as they were travelling to a concert along the Wye valley between Builth Wells and Brecon. They saw a group of children riding a pretty two-year-old filly. They stopped, enquired about the owner, and within a few hours they had bought the filly.

She was named Fortress and although she had no pedigree, she was certainly of Welsh parentage. So, she was inspected and recorded in the Stud Book as 'foundation stock'. It was a purchase that interested Barbara for more than one reason. The Brecon cob was different from the Cardiganshire type. It was larger and suitable for riding, the type of animal that Pauline and Barbara wished to introduce to their breeding programme. The Cardiganshire cob was heavier, bred mainly for harness work on farms, but the main characteristic apart from strength and power was their high stepping trotting qualities. The high action drew attention and applause at shows and sales, and they were often given colourful names such as Trotting Jack, Trotting Comet and High Stepping Gambler – stallions that were renowned for their power of movement and they always drew the crowds.

Llanarth Fortress had been registered as foundation stock in the Welsh Stud Book, and she was kept for riding and pleasure but she did produce one filly foal as a result of mating her with Llanarth Prince Roland, son of Brenin Gwalia and Llanarth Vega. Llanarth Rocket was born and she proved to be an enormous influence on the breeding programme by producing her one and only filly, the great Llanarth Flying Saucer.

Sometime later, Barbara returned to Breconshire to visit a well-known breeder, Gwilym Morris and she immediately picked out a glorious fox-red chestnut foal, the son of a well-known cob Pistyll Goldflake. She bought him just in time, as his tail would have been docked a week later. She registered him as Llanarth Goldcrest and he too became an influential stallion at the stud.

These purchases crystallised their thoughts and aims on the characteristics of the type of cob they would breed. They made two fundamental decisions. If the cob was to survive in postwar Wales, it would have to be multi-functional and adaptable, suitable for riding in addition to driving and harness work. They set their sights on a breeding programme which would, in time, develop both characteristics. The second decision concerned the docking of tails. It had long been the practice in the cob world to dock tails, but Pauline and Barbara thought it inhumane and unnecessary, since cobs and shire horses were no longer working horses. Long tails tended to become entangled in shafts and machinery, but since mechanisation and tractor power was supplanting the horse, they considered the custom should cease, allowing the cob to be shown in its full glory. They began a campaign to stop docking, much to the annoyance of the old Welsh breeders. 'Why change a practise at the request of new breeders,' they said, and worse, 'Why listen to crackpot ideas from women – long tails hide the strength of the hind quarters from the judges.' But Pauline, a commanding figure at meetings and gatherings rose above the scornful mutterings at sales, shows and markets, and used every means possible to persuade agricultural and political authorities that docking was out-dated and of no merit. The ladies of Llanarth simply refused to budge. They travelled around to lobby leading breeders and dealers, and one day called on Roscoe Lloyd, a well-known breeder who had found a regular market selling horses to dairies and breweries in London. Pauline's message was clear – docking was inhumane and unnecessary and with typical no-nonsense straight talking approach, he replied, 'We've got short tails and you've got long tails and that's it.' It seemed to end the discussion but Pauline was a past master at having the last word, 'You wait,' she said. The discussion continued on many occasions; they became good friends and soon with the help of Tom Herbert, the enlightened local veterinary surgeon, the campaign gained ground and by January 1st 1950, not only had they won the argument and increased interest, a Bill had been taken through Parliament and anti-docking for shire horses, cobs and ponies became law.

Barbara and Pauline were becoming recognised as serious

breeders and in 1948, Barbara was elected for two years to the council of the Welsh Pony and Cob Society, who were the guardians of recording, registering and categorising ponies and cobs. When the society was established in 1902, it was decided to divide the Stud Book into four sections, each one categorised by height. The first two, Sections A and B were Welsh ponies: A being under 12.2 hands high, B being over 12hh and under 13.2, and the two sections for cobs and cob types were C for 13.2hh and under 14.2hh, and D for over 14.2hh and under 15.2hh.

This was revised almost fifty years later when Barbara was invited to join the deliberations of a sub-committee to discuss the proposal to enlarge the scope of section B to include Welsh ponies of riding type. When they had completed their work, it seemed the height of animals remained more or less constant for the four different sections, but other breeding qualifications were introduced to clarify the need to introduce the riding type of pony within Section B.

'When Goldcrest became a two-year-old,' Barbara wrote many years later, 'he was placed with other young stock at Frongoch Farm. We also kept a couple of mares there, but they were kept well separated from the young stock. Our man at Frongoch never told us that one day Goldcrest had jumped over the banks and spent the day with the mares. In April, they both produced foals the same day. One of the mares, Kilda, was the daughter of the yellow dun cob stallion, Cardi Llwyd, and her foal was a pale fawn colour such as we had never seen. He was a palomino, not a type or a breed, but a colour.' Barbara became excited because she had grown up at Pentre riding the golden ponies with their attractive white manes and tails. Their origin at that time was a matter of conjecture and experiment and it led Barbara to become immersed in the mysteries of genetics and scientific principles. The unexpected birth of a foal of the distinctive colour sparked a detailed genetic study.

'What a thrill he gave us,' said Pauline of the foal. 'All the more because his arrival was unplanned. We called him Llanarth Braint there and then, in token of the honour that he had unexpectedly bestowed on us.' (Honour being the English translation of Braint.) The tailpiece to the story caused some amusement. The two mares who had enjoyed the day with Goldcrest at Frongoch had been taken to a local sale at Llanybydder the previous November, but because they hadn't reached the modest reserve price of £30 – bidding completed at £29 – Pauline and Barbara decided to withdraw and both mares were taken home. They wintered happily at Frongoch when fate or luck, or both, intervened in 1948, and Llanarth Braint, sired by Goldcrest with Kilda was born. He matured and although he was not immediately acceptable to the traditional Cardiganshire Welsh cob types, he went on to become known worldwide. The animosity towards Braint within the cob world grew. He was a palomino and to many that colour was a joke. His action was not of the Cardiganshire high stepping stride; he was more of the Breconshire type and although he was a handsome stallion who enjoyed playing to the gallery, if the judge was a Cardiganshire man, Braint stood no chance – he was rarely placed first in Welsh show rings. But Pauline and Barbara hid their disappointment and slowly pursued other avenues to make Llanarth Braint the most attractive Welsh cob at shows and events held in England and Scotland.

The postwar decade ended with another major change of fortune. The Llanarth stud, which included Welsh mountain ponies in addition to cobs, was increasing in numbers, and their general farming enterprise of cattle, sheep and pigs was expanding. They needed more land. Fortunately and coincidentally, two farms, Blaenwern and Gofynnach Fawr, which had been the heart of Enid's family estate at the turn of the century, came on the market. Enid was already the owner of her cottage, Rhydfawr, where she would often return, but at that time she had little interest in farming, as her teaching role at the Guildhall plus her involvement in concert work with The Dorian Trio was all consuming. Pauline and Barbara decided that the two farms would be ideal for their purpose and Barbara was encouraged by Pauline to make Blaenwern her Cardiganshire home. Her decision to make a further investment in the area came at an opportune time. The Attlee Labour Government, elected in 1945, imposed heavy taxes and death duties on land owners of large estates and many were sold and Barbara's family estate Pentre was no exception. The family, especially Barbara's brother, Arthur Owen, was under tremendous pressure and he eventually sold and Pentre mansion became an independent school. Her mother, well into her seventies was becoming frail, and Barbara invited her to move with her to Blaenwern, where she could give her care and comfort. It was a substantial farmhouse, square and solid. The core of the building was between four and five hundred years old but there had been many additions and renovations. Her mother must have been concerned at the course of events, leaving Pentre after forty years of life with servants in a large mansion, for a farmhouse in a very different community. But Barbara made sure the house was furnished with furniture and furnishings from Pentre, and soon it became a comfortable and elegant home.

CHAPTER FIVE

L LANARTH BRAINT RETAINED THE golden colour of a palomino with white mane and tail until he was a three-year-old. Then a darkening gene took effect, as it does with some chestnuts, and he became a liver chestnut but his distinctive mane and tail remained white. Fortunately, he did not pass on the darkening gene to any of his progeny and he went on to produce many excellent palominos. That was one of the reasons why the Cardiganshire cob fraternity did not fully accept him. But it was also because of his conformation with more of the Breconshire type, a bigger animal and a riding type. He was a truly magnificent stallion with a fine head, strong and a free-flowing action, and a wonderful temperament. He knew instinctively how to attract attention in the show ring. Barbara wrote, 'We showed Braint as a two-year-old at the Royal Show and he got second prize. He was already a spectacular mover but I think the cob judges did not approve of his long tail or the palomino colour.'

This became a long running argument with other breeders. Barbara and Pauline were not only introducing the characteristics of the larger, riding type cob of Breconshire into the Cardigan type, but now they were talking of their intention to breed more palominos. It was heresy; it was direct meddling with the inherited characteristics of a native breed. Cardiganshire cobs had traditionally been coloured either chestnut or black, but Barbara and Pauline held their opinions. Barbara, who had always been interested in the science of genetics, now became immersed in the subject until she had mastered its mysteries. Breeding palominos became an absorbing passion.

Palomino is a colour, not a breed or a type, but the colour requirements are strict for classification. It occurs in many native breeds of the world and in Wales it is a colour found in the Welsh mountain pony and the cob. It can be traced back through many dun coloured cobs to Cymro Llwyd, since the yellow dun with black mane and tail carries the same genetic factor as the palomino, and if mated with a chestnut will often produce the true palomino colour.

In Europe and America the colour originates mostly from Spanish horses, often called Isabellas, in honour of Queen Isabella of Spain who greatly esteemed them. Interestingly ,the colour does not occur naturally in either the pure thoroughbred or the Arab but palominos have been crossed and recrossed with those breeds so that many carry over 90 per cent of the blood in them. One of the most famous palomino horses at that time was Trigger, acknowledged as 'the smartest horse in the movies'. He was, after all, the Wonder horse and the faithful mount of the cowboy star, Roy Rogers during the 1940s and 50s and together they made dozens of Westerns chasing and thwarting the bad guys to serve peace and justice. As Roy Rogers said of him, 'He could turn on a dime and give you some change'.

The golden colour inevitably draws attention at shows and parades and when they seriously began showing them in the early 1950s the Llanarth palominos were much in demand. In 1956 Barbara wrote a long article explaining in detail the basic principles of breeding for colour. She wrote:

> The pleasure of observation and cross-breeding can be greatly increased by a knowledge of the basic principles of inheritance, and we are especially well favoured in this country by the variety of colour to be found in some of our native breeds of ponies. Genes, the units that carry the inherited characters are always present in pairs, one inherited from each parent. Sometimes the genes in the pair are similar, in which case, the animal is pure for that colour; sometimes they are dissimilar making the animal impure. Some genes are dominant and will be sufficient to produce a colour, but recessive genes must be present as a pair to produce a characteristic colour. Black and chestnut are the dominant foundation colours

of the horse. Black is dominant to the chestnut, and chestnut is recessive to black.

The intricacies of the crossbreeding process and a deep knowledge of the characteristic and behaviour of certain genes gave her work greater emphasis as they developed characteristics and type for the Llanarth stud. One fact soon became apparent. Crossing a palomino with a palomino will only give a 50 per cent chance of a palomino foal with a 25 per cent chestnut and 25 per cent chance of a cremello. The reason for the golden colour is the result of another gene, a dilution factor.

She continues in her article on breeding palominos:

> The diluting gene acting on chestnut produces the palomino colour. Since so many blacks and bays carry chestnut recessively, many palominos are produced by crossing the palomino with a Bay, and a palomino with Black. A double dose of the dilution factor mating with a chestnut produces light cream with pink skin and blue eyes known as albino creams or cremellos.

The genetics of colour is complex and Barbara read widely and studied wisely. She chronicled her findings and jottings, writing with a pencil on every bit of spare paper she could find at Blaenwern. Many of her notes were written on scraps of paper – occasionally on the reverse side of complimentary slips sent to Pauline by Gwilym Edwards, Secretary of the Welsh Black Cattle Society. She made countless lists of dominant and recessive genes, alleles, diluting factors, the mathematics of genetics and the possibilities of pigmentation as a result of crossing certain colours. She recorded a long list of possible names for palominos male and female; interestingly every name ending with the letters 'el'. Over twenty had been used when I looked at one early list, including Gabriel, Petrel, Joel, Manuel, Samuel and Daniel. There was a list of Welsh names for reference to be used for all types of animals, from Cymro,

Gelert, Pryderi to Rhosyn, Dilys and Ceri. Barbara's thorough scientific research over fifty years, coupled with its practical application enhanced the wealth of knowledge she achieved as a breeder.

The stud expanded in the early 1950s and three or four of their stallions had been given a premium from the Welsh Pony and Cob Society and requests from other breeders would arrive at Blaenwern for their stallions to mate their chosen mares. The old practise of travelling stallions from farm to farm, a specific route every year from April to June was aptly called 'the season', but by the end of the 1940s, lorries and landrovers had displaced the smartly dressed horsemen walking the road. It was the custom to advertise the merit of each stallion on small cards which were sent to other breeders – the language being colourful and flamboyant. This was how David Rees promoted his prized stallion, Brenin Gwalia, in 1943:

> He will serve a limited number of mares at £2/2 each. He is a blazed-faced chestnut with four white limbs and flaxen mane and tail, stands 14.3 hands high and goes great guns all round.

Both Brenin Gwalia's grand sires were exported as equine pioneers of Welsh cob breeding in two continents; Trotting Gambler went to Australia where he triumphed in leading shows and High Stepping Gambler went to South Africa, where he was serving mares at £10/10 each. David Rees went on to say, 'If anything happens to Brenin Gwalia, a substitute will be provided and must be accepted.'

Barbara and Pauline bought an old lorry with a canvas top to transport their stallions around farms. It was no mean job to load and unload strong, excited, frisky stallions and drive them along the narrow twisting Cardiganshire lanes and then to handle and control the sexual needs of prancing stallions when they saw an attractive 'on heat' mare. The old breeders found these farmyard scenes of procreation handled by two women quite embarrassing and totally unacceptable. It was unheard of but, as William Lloyd of the Geler stud, a well-known breeder and judge explained, 'Farmers laughed at two women handling a stallion at such occasions. It was a joke for the men. But my mother for instance, when she knew a stallion and mare were at their business in the yard, she would shout at me from the kitchen 'Shut that door'.'

Ifor Lloyd, of the famous Derwen stud high above Cardigan Bay remembers as a young boy the sight of the two ladies bringing a stallion to the yard, 'It was a novelty. This was a man's world in the true sense of the word. And a Cardiganshire man's world at that.' Wynne Davies recalls attitudes of the time, 'It was a joke at the beginning, it couldn't be taken seriously, but their successes soon turned scorn and laughter to admiration for their astuteness and courage.'

There was always keen competition for stallions to be awarded the 'Premium' which in the cob world was interpreted as a 'bonus' or a financial subsidy. It came about when the Board of Agriculture made annual grants of money to be used as bonuses or awards for a number of Welsh stallions selected by a breed society such as the Welsh Cob Society.

The stud engaged a young girl, Anne Lyne, to assist with schooling, training, leading and riding their ponies and cobs. She was the daughter of a mutual friend and she knew how to feed a horse and was a good rider who could run and keep up with a Welsh cob trotting at full stretch. Barbara wrote:

> She had been well-trained by the local rat catcher and she quickly became our stud groom. She was a superb horsewoman who had a thorough understanding of how to get the best out of them and how to show the horses to advantage, whether riding or leading at shows and sales... I remember Mrs Yeomans, who did a lot of announcing

at shows saying over the loudspeaker at one of the English shows,
'In Wales they need two men to show a cob stallion. Here you have
one young girl.'

Their reputation as breeders grew and the cob community
began to accept that the ladies of Llanarth 'knew what they
were talking about'.

Blaenwern became the core of their cob and farming
enterprise. It became the home of the Llanarth stud with
Barbara and her mother living in the house while Pauline
remained at Bwthyn. They were a complimentary
partnership. Pauline was the bohemian: robust, noisy,
outspoken, persuasive, independent, hard-working and an
enthusiast with a liking for the simple life. She preferred
being out with her animals rather than in the house cleaning
and cooking. Barbara had engaged a maid to run the house
and she was a generous benefactor to the whole enterprise,
which her mother regarded with suspicion; she had forfeited
a wealthy lifestyle for the satisfaction of research and its
practical application. She always dressed the part of a
country woman and was never without her flat green beret
and her corduroy trousers. In truth, she was not a particularly
approachable person and occasionally she was rather aloof.

Carole Knowles-Pfeiffer, who was an agricultural student
at University College of Wales, Aberystwyth at the time, was
sent to Blaenwern 'to gain one year of practical experience
for her course.' She was not a cob enthusiast but she assisted
Pauline with all aspects of the farm, especially cattle, pigs
and chickens. She lived in the house with Barbara and her
mother:

> ...who was very stern, wanted everything done properly... It
> was posh with everything in its place. Barbara's mother was the
> matriarch. She had a maid and a nurse. Mind, it was fun looking
> back, although I was not a cob person (I didn't handle them), but I
> had respect for the work. Barbara was cool, a reserved person and

she put the breeding programme on the rails. She wrote an article on colour in horses – it was well-researched and inspired. She had a wonderfully analytical mind. The material in that article has stood the test of time.

Unlike her mother, Barbara was a hoarder of books, magazines and papers and all manner of bits and pieces she saved from the weekly detritus. Anne Wheatcroft, a close friend, recalls one of Barbara's traits that used to drive Pauline:

> ...up the wall. In Blaenwern they used to take the rubbish bins down the lane to the road. If Pauline did that job, Barbara would go down the road a little later, go through the rubbish, pick out what she wanted to keep, especially binder twine – she could never throw away the smallest bit of twine.

When asked about Pauline, Carole replied:

> She was clever, direct and slightly dotty... In fact, we were a right bunch in that house. I think we were all slightly loopy, but it was fun. Radio 3 music all day. Music in the evenings when we'd all dress up. Dogs and cats everywhere and behind it all the serious business of cob breeding and animal husbandry.

Len Bigley pointed out that Pauline was the only one of the three, Enid, Barbara and Pauline who was not Welsh. But as everyone remembers, she was interested in all things Welsh – the characters she met regularly at farming events and especially being part of the Welsh community. Blaenwern farm was a triple enterprise of Welsh cobs, Welsh Black cattle and Welsh pigs. In 1953, a group of Welsh Black cattle established by Pauline won the coveted Prince of Wales Cup at the Royal Welsh Show. The herd had grown from ten to eighty and included twenty milking cows and ten 'nurse' cows. When a greater variety of feeding stuff became available after the war years, they raised their milk production from 600 gallons per cow to an average of 900. In the same year, they won a cup for milk records and inspection for all breeds in the county. Pauline believed the importance of breeding the Welsh dual-purpose type of cattle because, 'They are docile,' she said at the time, 'They can look after themselves; the roughest and wettest of weather leaves them unperturbed and they will pick a living from the poorest grazing.' The Welsh Black cattle were bred for their meat in addition to milk production and were known as a dual-purpose breed. In an undated pamphlet signed by them, they wrote of the history and the qualities of the Welsh Blacks and answered the question of why so little had been heard of them in comparison to other breeds:

> The isolated Welsh hill farmer had little inclination or chance to push his cattle beyond his immediate neighbourhood – though the value of the beef store cattle was long ago appreciated by the graziers of the Midlands and eastern counties. But in Wales there was little communication between north and south so that two distinct types of cattle developed – sturdy, small, compact type adapted to the mountains in the north and the larger and heavier beasts known as the Castle Martin in the south west. The good points of both are apparent in the present breed and the leading show animals of recent years have been the result of wise and far-sighted mingling of the two strains.

I'm sure the pamphlet was their contribution to promote the breed in other markets.

They had also bought wisely to develop the foundation of their Welsh pig enterprise. Their main aim was to preserve the true character and vigour of the breed without having to cross with other types such as the Landrace which was increasing in popularity. Two of their sows qualified for the Improved Advanced Register and many of the progeny were sold to other farms for breeding – the rest were sold as weaners or kept to fatten for bacon or pork. Since the war the Welsh breed had prospered because of its fast weight gain, hardiness and its ability to thrive in a variety of conditions.

Production costs were kept down by careful management and improvement of grassland, much of it ploughed and planted in new leys of Aberystwyth strains of grass. Careful use of fertilisers, very little nitrogenous mixtures, but mainly basic slag was used to promote growth coupled with a liberal spread of farmyard manure in early spring. The traditional winter feed for a mixed farm was hay, oats and the traditional Siprys, a mixture of oats, barley and corn. The arable acreage was supplemented with kale and roots for cattle and sheep.

By the mid-1950s their estate had grown to 300 acres which included the smallholdings, Frongoch and Dolau, to the larger units of Gilfachreda and Blaenwern. To operate the farm effectively, they engaged a labour force of three: John Hockaday in charge of the rearing farm at Frongoch was a knowledgeable stockman, Abel Jones at Dolau who could turn his hand to anything and Ifan Maclaren, a Llanarth man of Scottish descent who had the reputation of being a master thatcher and at harvest time was the best stack and rick builder in the area. Pauline always recognised their talents and their contribution, 'They have been invaluable and we often feel they get too little credit for our success.' They understood and supported the development and quality of native breeds of cobs, ponies, cattle and pigs. It was a collective mission. Pauline and Barbara had very high standards and they rejected anything that did not match those standards. In the early 1950s they set their goals and ambitions at a high level but the market was limited because unfortunately few people outside Wales realised or understood the merit and quality of Welsh native breeds. To rectify that particular problem the answer was plainly in their own hands. It was imperative to put the emphasis on promotion and to show breeders outside Wales the quality and distinctiveness of their product.

CHAPTER SIX

L LANARTH BRAINT WAS THE perfect animal for promoting and marketing the quality of the cobs bred at the Llanarth stud. He was a spectacular mover and an attractive colour but, at his first showing at the Royal Welsh Show, the judge did not approve of undocked tails or the palomino colour. Braint was placed second. Barbara felt they should extend their activities beyond the claustrophobic Welsh cob world and show their animals at showcase events in other parts of the UK. It was a brave decision and, as Anne Wheatcroft, a well-known breeder who in the 1950s became a close friend, said, 'I remember them as great ambassadors for the breed and were notable visionaries, well before their time.'

The Llanarth message was simple. The emphasis would change from the cob as a work horse, docile and strong, to being suitable and comfortable for riding and light harness work. There was scepticism in many quarters. Barbara wrote of another incident at the Royal Welsh Show when eventually, following persistent lobbying, the society included a new riding class for cobs in the show schedule:

> The judge was an Englishman, a judge of hunters and hacks. He pulled Braint in first in a class of about a dozen, but when he had to ride him it was obvious to those watching around the ring that he was scared stiff. He had only seen Welsh cob stallions tearing around the show ring, with heavy shoes and a long leading rain. 'Is he alright?' he asked Anne, our groom. She nodded. We knew Braint was a lovely ride. The judge mounted rather gingerly and moved off carefully. Braint walked quietly. He gently urged him to trot, then to canter and finally to gallop and he came back to the line beaming with pleasure. Braint won the class.

There remained stiff resistance to the stance taken by Pauline and Barbara, and they would hear the comment,

'Cobs aren't made to canter,' and Pauline would emphatically reply, 'If you want to sell your cob, you must, MUST ride them,' knowing that the carrot of financial gain would eventually appeal to a Cardi.

Today Wynne Davies acknowledges that the derision they encountered from other breeders would have irritated and exasperated Pauline and Barbara, but at shows in England they encountered praise and success. 'Llanarth Braint was a joke in Welsh show rings but he impressed and drew applause from judges and breeders in England.' They travelled extensively from Harrogate to Ascot, from Edinburgh to the Royal Show in the heart of England, and soon the Welsh cob and the Llanarth stud was becoming successful and widely known. Wynne Davies remembers:

> When a class for cobs under saddle was first introduced at the Royal Welsh Show, a leading Welsh breeder decided to enter one of his best stallions. He wore his trilby hat, his Sunday best jacket and trousers with the stirrups on his saddle pulled high, resembling the style and action of a racing jockey, but he could only sit sedately as his cob walked quietly around the ring. In fact he could do little else. Trotting would have given him an uncomfortable bumpy ride given the stallion's high action, and cantering was out of the question – he had never been trained. Llanarth Braint came in, obviously enjoying himself. Strong and handsome, he was a wonderful advert for the breed.

His success rate increased as his new handler in the show ring, Hefin Davies of Talgarreg, developed a particular empathy and mutual respect. Hefin Davies could run – really run – setting his pace to Braint's pace and allowing him to show-off and to be a vision of muscular grace and strength. At Harrogate in 1958 at the Ponies of Britain Show, he won the championship in hand, the first prize under saddle, another first in harness and the Supreme Championship of the whole show. As the years passed he was often competing against his progeny. He occasionally stood second to his sons but his greatest 'moment of glory' came in 1969 at the age of nineteen, when he put on the show of his life to stand Champion Welsh cob followed by his descendents of all ages and sexes.

But despite all his success Braint remained unpopular with the traditional Cardiganshire breeders. He contained too many of the Breconshire cob characteristics and the palomino colourisation set him apart from the 'home-bred' black or dun. But, Barbara and Pauline stuck to their policies and as Ifor Lloyd remembers, 'The old boys poked a lot of earthy fun at them, but these two – they didn't care a jot, they argued their case and carried on.'

That first decade of business, the 1950s, was one of tremendous achievement for the stud and demand for their cobs at home and abroad, especially the palominos, increased year-on-year. Prices at sales were escalating, a four-year-old Welsh palomino pony of cob type, Llanarth Pimpernel, a Section A stallion, was sold Lot 159 at the 1956 Fayre Oakes sale for 100 guineas and there was also a marked increase in exports. Keen breeders from the USA bought the attractive Welsh mountain ponies which were eminently suitable for children as riding ponies, whereas large cobs appealed to Argentinian farmers to cross with the native 'gaucho' ponies. By the mid-1950s Llanarth Braint had sired many palomino foals from chestnut and roan mares and a most profitable business grew into the largest stud of palominos in the UK. Billy Smart and Bertram Mills, leading circus owners, took an interest. They were famous names in the circus world, often travelling worldwide searching for ponies of the right colour, stamina and temperament and when they visited Blaenwern they were not disappointed. The unusual palomino colour of ponies and cobs, plus the docility, temperament and intelligence of the animals and their matching strong body conformation, were an appealing proposition as replacements for the familiar greys and chestnuts of the circus rings at home and abroad. Many of the palominos were Welsh part-breds, horses registered with a minimum of 12.5 per cent Welsh blood in them, the result of crossbreeding. There was a growing demand for horses to perform as show jumpers, driving, dressage and eventing, or in the circus ring. With Barbara's skill for selecting the right stallions and mares, it gave the progeny different characteristics, strength and docility, or an attractive colour, flowing movement and action with a temperament which could be trained to adapt to the confines of a circus ring.

A group of four palomino ponies left Llanarth one fine autumn morning in 1954 to catch the train from Aberaeron station bound for Zurich in Switzerland to join the Swiss National Circus – another indication of their growing export trade. The appeal of the Welsh cob was growing and the export trade increasing, especially to Europe and the USA. Buyers and breeders would come from far and wide in search of good bloodlines. As the largest palomino centre in the UK, the Llanarth stud had become a lucrative business and by 1960, the Welsh Pony and Cob Society formally recognised the Welsh part-bred and it was incorporated in the Stud Book.

If Llanarth Braint was the turning point in the development of the Llanarth stud, then Fortress, their first purchase, has also a special place in its early history. She was mated with Llanarth Prince Roland and produced a filly which they named Llanarth Rocket. It was to be Fortress' only filly foal and because she was of foundation stock, her colt foals could not be registered. But when Rocket was three years old, she was mated with Braint to produce another champion mare, Llanarth Flying Saucer, and it was she who had an enormous influence on the Welsh cob breed. Between 1954 and 1976, she produced nineteen foals; all but one was sold for breeding, showing or export. The one stallion, Llanarth Meteor, was retained at Blaenwern.

When they wanted to improve or to develop another line or characteristic in their cobs, they bought wisely. A mare named Rhosfarch Morwenna, owned by Cardiganshire breeder Richard Tudor, was such a purchase. She was a beautiful animal with a long reach of neck and when mated with Braint, she produced a wonderful series of eleven champions: Meredith ap Braint, Marc ap Braint who became a top sire in France, Math ap Braint and Maldwyn ap Braint who became top sires in Canada in the 1970s.

Pauline was to remark, 'It has been a revelation to many outside this country that such an animal as the Welsh cob has persisted through the centuries.' At the end of the 1950s the stud was well-established and they were accepted as breeders of quality animals. Both had become members of the Welsh Pony and Cob Society in 1943 and as their reputation grew, Barbara was invited to join a society sub-committee to deliberate on the need to establish a section for Welsh ponies of riding type, or to incorporate the quality within one of the breed sections A, B, C or D. The committee met in September 1947 and four and a half months later, at the Annual General Meeting, all their recommendations for Section C were accepted – the first step that the society had taken to accommodate the pressure that Barbara and Pauline had put on other cob breeders. This type of cob became eminently suitable for pony trekking centres as they grew in popularity in Wales. Barbara was elected a member of the Council and Pauline became an active member of the Welsh Black Cattle Society.

Pauline was in her element at shows and sales. She liked people and there were very few she didn't have time for and even if she'd had a fierce argument with someone, it was immediately forgotten. Len Bigley, who was at the stud for twenty years added his recollection, 'I can only remember

Teleri Bevan

two people she didn't trust.' William Lloyd remembers an occasion when his father, John Lloyd, met Pauline and Barbara at the Lampeter show. 'A fair crowd stood around admiring Llanarth Braint when Pauline challenged him, 'You see Mr Lloyd, they admire Braint, he's such a lovely horse, all the people have come here just to have a look at him.' 'Oh', replied my father, 'I'm sorry to disappoint you, but I've come to see the owners more than the stallion – you're far more interesting than what he is."

Once a year they took to the hills and into the heart of Welsh mountain pony country. They wrote of their first experience of an annual sale of mares and foals gathered from the Cambrian hills at a lonely spot near an isolated inn not named in the article:

> Mares and foals jostle one another packed tight in a yard. More are arriving in groups rounded up by boys. Some are driven in to sheep pens nearby, others into a little plantation, some are left in the lane, greys, bays, roans, chestnuts, duns, blacks and the occasional palomino – a kaleidoscope of colour. And the human types too – old hill farmers, shepherds, dealers, boys with a newly broken pony to sell, the enthusiast looking out for a smart colt to show the following year and many well-established breeders. It is a meeting place full of good natured talk stimulated by the free flow of beer. As we lean against the yard wall, the auctioneer approaches and asks if we will act as judges and award prizes.

Judging takes place before the auction and is not an easy task with over a hundred foals to judge who are scattered in different places, scared out of their wits:

> We go first to the yard, push through a jostling mass, pick out a good-looking foal and hastily two young lads grab it and coax it into a ramshackle shed. We pick out about ten. Next, the sheep pens, followed by those in the plantation and the lane. Poor little mites, they are carried or dragged to join the others. We consider each one and after careful weeding out there remain three little greys of real quality. We award the cards and head for the beer and refreshments.

69

> The auction follows until dusk. As the light fails, the little mares are driven back to their native hills whinnying anxiously for their foals which they will never see again.

The importance of such sales and the gathering and selling of ponies that roam wild on the mid-Wales Cambrian hills, where only the best have survived nature's rigorous selection, has been the foundation of quality and characteristics within the breed:

> Man has done what he can during the last fifty years to choose the best stallions for premiums and removing unwanted ones, but now many beautiful mountain ponies are bred in the luxury of the lowlands where stamina and other qualities may decline and be lost. It is up to the hill farmer to maintain the native stock and to see that the breed loses nothing of its hardiness and intelligence.

Fifty years ago they had looked to the future and voiced a real concern. Today a recent report by the Rare Breeds Survival Trust categorised a list of rare breeds of horses which included the Cleveland and Suffolk Punch as critical, the Fell and Exmoor ponies as endangered, and for the first time, the Welsh mountain pony had joined the Dartmoor and Highland ponies as being vulnerable.

Horse fairs were an important feature in the days of horse power and there were three notable ones within the area known as 'Cob Country'. I have no doubt that the one that Pauline describes was held at Ffair-rhos, on the foothills of the Cambrian Mountains, near the Teifi Lakes. It was an autumn fair (October 23), the last of the season when ponies and foals were sold after their summer on the hills. Spring fairs for selling working horses included Ffair Garon (March 16) held in Tregaron, and Ffair Dalis (May 7) in Lampeter where buyers came to buy for merchants and industries in towns and cities. But these dates had an agricultural significance in addition to trading horse flesh. Customs grew around the farming fair calendar. March 16 was the date to sow oats and as the days became longer, the date to stop lighting candles and lamps; Ffair Dalis signified the time to sow barley, and Ffair-rhos was the time to light candles and lamps once again as the number of daylight hours became fewer.

Such events as fairs and sales, part work but mainly as meeting places and social discourse, made their lives at Blaenwern full and stimulating. The transfer of news and gossip was central to these occasions. There were other days when, at the end of a hard day's manual work, the walls of the old longhouse house by the stream would echo to the sound of music, especially when Enid returned for weekend breaks or holidays at Rhydfawr. She had transformed her cottage into an elegantly furnished haven – her grand piano was housed in the music room which she had converted from the cowshed. It was here, hidden away in the woods, that they would often retreat to their music-making or, occasionally, they would invite friends to join them for musical evenings at Blaenwern where Barbara and Enid at two grand pianos would join the violin and cello. During the day, the yard echoed with the sound of Radio 3 as they went about their work, especially with the horses. Music remained an integral part of their lives, and was a balm, a counter to the derision they encountered during the first years of establishing the farm. They would travel to London to attend recitals and concerts and stay the night at their club. In addition, Enid and Pauline maintained their full schedule of school concerts and lectures as The Dorian Trio during the 1950s when the remuneration for a school recital in south-east London was £2.12.6 per artist. When 1960 dawned, that work came to a natural end. They had given almost forty years service to musical education, not only in Wales, but at schools and community centres in the southern counties of the UK. Thousands of children had attended their concerts, many were inspired and others, who were not as committed to classical music, would never forget the sight and sound of The Dorian Trio.

Pauline and Enid were well into their sixties, Enid sixty-nine, Pauline sixty-four and Barbara ten years younger at fifty-three. Although Enid was well passed the official retirement age, she had been elevated to a professorial role at the Guildhall and she decided to remain teaching and working with gifted young students, 'as long as they want me'. Her lifestyle was rich and varied; she lived in an elegantly furnished flat in Regent's Park. Her niece Jane, who now lives in the USA, remembers her as a superb cook, entertaining her and her brother Edward to sumptuous teas. One day, they found her searching for a festive summer punch in her copy of *Mrs Beeton's* for Edward's twenty-first birthday party. The book was a gift from her parents when she left Pontypridd for Bangor, and interestingly every page dealing with alcohol had been carefully removed:

> I still have a Georgian silver christening mug (a little odd for a Unitarian's gift) and I daily use the silver napkin ring with Jane engraved on it. I was very fond of Enid who always seemed exotic in my eyes – long strings of lapis lazuli that bumped gently on her bosom as she moved, *pince-nez* on her nose – so popular and fashionable in the nineteenth century but by the 1940s and 50s worn only by the elderly – and the garlands of laurel leaves and holly berries she threaded at Christmas.

Pauline's life was not without its family loyalties and responsibilities. She was not a cook, not really interested in clothes and positively disliked housework, but she liked and appreciated 'nice things'. Pauline had a wide range of interests – music, the arts and the environment. But, at heart, she was an academic. She had edited and published the third edition of her father's volume of poems, *The Door Keeper*, which had been out of print for many years. The constant demand for some of the poems and unpublished works which came to light in his papers convinced her that a further edition would be welcomed. It was a labour of love and gave her immense satisfaction. The divorce of her eldest sister Alice in 1951 was a cause for much family concern. Her marriage to the eminent psychoanalyst and paediatrician, Donald Woods Winnicott, lasted for twenty-eight years and although they were initially happy – she was beautiful and artistic – their marriage was never consummated. Donald was sexually impotent and infertile and this put a strain on their relationship. He immersed himself in his work with children at Paddington Green Children's Hospital and Alice, a skilful and knowledgeable mineralogist, went to work on ceramics and glazes at the National Physical Laboratory, and later became the manager of a pottery in Kent. Sadly as years passed Alice developed a mental illness. Her nephew recalls, 'She was often after that a bit curious, but she ended up managing a pottery so she can't have been too dippy.' But as time went on her condition deepened and she became a disturbed and an unhappy woman. She rarely bathed and used to commune with the spirit of novelist and critic D H Lawrence through her pet parrot. Donald moved to work at Oxford during the war and it was there he met Clare Britton, a pioneering social worker developing services for deprived children. Within a short time, their professional relationship turned into a personal one. They married two weeks following his divorce from Alice and became, 'Two people who delighted each other and delighted their friends.'

Alice remained in the south of England but her erratic behaviour was causing concern, although her work as a painter and potter attracted attention and was selling well. She often signed her paintings 'Claverdon', although the name was not of particular significance to anyone in the family. But, an increasingly worried Pauline felt her sister could not be left to fend entirely on her own, so she suggested that she came to live in Cardiganshire and to buy Gilfachreda, the first farmhouse Barbara had bought in Llanarth when they set up the partnership. The transaction completed, Alice moved to the Welsh countryside to continue making beautifully designed pots and to paint landscapes of Wales, but sadly, she was always known locally as 'the odd lady'.

Barbara had been in partnership with Pauline for seventeen years. Her mother had become very frail and life was becoming more and more difficult with her needing constant nursing and home care. Her mother had not found life at Blaenwern congenial; she was convinced that Pauline, with her forthright attitude and lacking the finesse of a true landowning lady, had turned into a bully, expecting Barbara to finance their activities as and when further investment became necessary.

But no-one today can offer an informed reason or accurately pinpoint why, in 1961, Barbara decided to leave Blaenwern. The business was prospering. Against all the odds, the stud of cobs and ponies, the Welsh Black cattle and pigs were in demand, and there was much more to achieve. Barbara's own recollections, in undated notes which she wrote sometime in her old age, suggest it was her mother's health that had been uppermost in her mind, 'I left Llanarth in 1961. My mother was an invalid and two friends in London who were buying a house in St John's Wood offered to share it with me. One of them was an excellent nurse and had become very fond of my mother and offered to look after her, so I migrated to London.'

CHAPTER SEVEN

PAULINE WAS STUNNED WHEN she first heard the news. They were very different personalities and their partnership of seventeen years, if symbiotic, had not always been harmonious, but it had certainly been productive and visionary. Barbara's scientific knowledge and practical application had been the key. She was the 'brains' behind the Welsh cob enterprise and Pauline had always recognised privately and publicly Barbara's unique input to the success of the stud's breeding programme and its development. From their first meetings Pauline had set out her vision and Barbara had possessed the skill and tenacity to convert the dream into a substantive reality.

The decision to leave had not been taken lightly. It was the culmination of an objective and logical process of evaluation from a combination of factors – her mother's illness and frailty, her own desire to return to academic studies and also, after seventeen years of hard, heavy work in grime and dirt in all weathers wearing unflattering dungarees and shirts, she just longed to dress in elegant fashionable clothes and to experience again London's cultural life. There was one quality that Pauline immediately recognised, once Barbara had voiced her decision, it would not be reversed. It was not to be a matter for discussion or persuasion. Barbara was fully aware of the practical difficulties and problems facing Pauline when she left. They had always shared the day-to-day workload of caring for all their animals – over sixty Welsh Black cattle including suckling calves, five pedigree Welsh breeding sows and between eighty and ninety Welsh cobs and ponies including foals. It was a daunting prospect for Pauline in addition to many, many other responsibilities and commitments. The most pressing would be financial. Barbara had always provided the cushion of financial security but

Pauline had little capital to invest. The intention was to sell the Blaenwern house and other dwellings on the estate and now the immediate problem was finding a buyer.

Initially, Barbara suggested they turned to their friend Anne Wheatcroft who lived with her mother in Gloucestershire and had developed the Sydenham stud of small Welsh riding ponies in the 1950s. She read an article about the growing popularity of the Welsh cob and, at one of the agricultural shows, she had been very taken with the strength and beauty of Llanarth Braint. Immediately she wrote to Blaenwern. Barbara replied offering her advice and guidance and a close friendship began, more so with Barbara than with Pauline. Anne's interest in the breed flourished and over the years she visited them many, many times, staying at the Feathers Hotel at Aberaeron. But, that one visit in 1961 turned out to be a critical time at Blaenwern. Anne takes up the story:

> They both asked or invited me whether I would please consider running the stud for them. After many discussions I turned it down and I'm very glad I came to that decision. For many reasons really, but mainly, I knew, I couldn't work with Pauline. However much I liked and admired her, which I did, I could never work alongside her. I wanted desperately to move to Wales – I wanted to move from where we were. They put a lot of pressure on me, they really did, but in the end I knew I couldn't get along with her.

Anne had too much respect for Pauline to explain fully the reason for her refusal. As it happened, Pauline had come to the same conclusion as Anne, she simply said 'no' in her forthright, stubborn way. The situation was fraught. Privately Pauline shrugged her shoulders, inwardly baffled and bewildered, attempting to hide her inner turmoil at the prospect of facing a bleak future. She was sixty-four and at a time in life when she could have believed that they could ease up on the range of their activities. Together, she and Barbara had staged their last major event together, an Open Day in the early autumn of 1960 to promote the cob sections C and D – an event which proved to be attractive and successful. The preparations had taken months and Barbara reported, 'The presentation was informative and everyone who attended the event was interested. One feature that seemed to impress was the blending of the cob and the pony to evolve the rideable quality we see today.'

Sometime earlier, Pauline had put forward the thought of holding an annual auction of cobs and Welsh ponies of cob type at Blaenwern in order to show the growing vibrancy and value of that section of the breed and to reduce the number of horses at Blaenwern. Barbara's reaction was a negative one; she was firmly against holding an auction sale on the farm and this without doubt became one of the major contentious issues between them. She didn't approve and although she and Pauline frequently argued and disagreed, this proposal caused more enmity than most. Anne Wheatcroft believes this was one of the major issues which led to the rift. Years later, Barbara admitted that she had been wrong, very wrong, because Pauline's long cherished ambition of staging the first autumn sale on the farm in 1964 proved to be a success. There does not appear to have been any sign of anger or animosity, but the accumulation of all the irritations from seventeen years of working in business together – the untidiness, the rudeness, the domineering attitude and the lack of appreciation and gratitude for the financial security Barbara had provided, had given her the impetus to end the partnership. Pauline could only explain Barbara's action with the comment, 'She's half American. And a typical American, and like all Americans she's had a seventeen-year-itch.'

Barbara wanted to move on. Her break with Blaenwern was complete and she did not return for three years, and that merely for a short visit. They would telephone and they exchanged Christmas cards which Pauline signed, 'with love, Paul'. These were cards used to advertise Llanarth premium stallions for potential breeders with a photograph of one of the Premium stallions as a frontispiece. Barbara set up home with two friends in a large house in St John's Wood, but her experience as a Welsh cob judge was often called upon by cob and pony societies, 'I did quite a bit of Welsh cob judging at shows and events in England,' she wrote, 'And also inspecting for the Palomino Society but I gradually lost touch with the cob world where I had been a council member for many years.' London life was appealing, she could dress smartly again, she went to concerts and theatres and she began going to libraries to take up more intellectual pursuits.

She had few regrets, and it seems she was satisfied that this was the end of a chapter in her life. Within months, the manager of the Anthroposophical Society Group operating a lending library in Museum Street, Holborn suffered a stroke, and Barbara was asked to take her place. This was a society she had joined in 1926 because of her interest and admiration for the philosophy and work of the founder, Rudolph Steiner. He had coined the word anthroposophy, from two Greek words, 'anthropos' for teaching and 'sophia' being wisdom, as a suitable title for his philosophy. It was based on creating a spiritual path to freedom and included scientific studies and research into subjects such as education, agriculture, medicine, architecture and many others which would lead to a better understanding of the spiritual nature of humanity and the universe. Barbara was to spend the next seventeen years, until her retirement, working for the organisation. She became the librarian at Emerson College, Forest Row in Sussex, an international centre for adult education based on anthroposophy. It was here, during her own research and studies, that she translated from the German with Katherine Castelliz, *Lebenstyfen der Erde* by Walter Cloos – *The Living Earth: The Organic Origin of Rocks and Minerals*. She became keenly interested in the principles of bio-dynamic farming. Rudolph Steiner was the pioneer of this system and made the first compilation of organic farming techniques. Barbara had often voiced her disapproval of using inorganic fertilisers on the land at Blaenwern but Pauline, in her usual forthright manner, pooh-poohed organic practises when the authorities called on farmers to produce more and more from the land, after the war. And there were occasions, in the first raw months of being alone when Pauline was at her lowest ebb, that she would vent her spleen with observations that Barbara had been taken in by the Anthroposophical Society because she was so wealthy.

Pauline came to rely more and more on the kindness of friends such as Betty Davies, who had come to farm in the vicinity and who was probably to become her closest friend in those first few months on her own. Many years later Betty Davies recalled her impressions. 'When Barbara left, Pauline needed all her courage, strength of character and faith to carry through the first gruelling years of problems.'

Pauline was determined not to give up however grim the future. She turned to Enid for advice and help, and in her direct and forthright manner did not mince words about her financial predicament. She knew the farm would not make enough money to invest in improvements and she had no capital. Banks and finance houses resisted and stonewalled requests for loans from women who were in business on their own, however successful they were. Within days Enid had suggested that she would consider buying back her grandmother's home, where she had spent so many happy childhood holidays. Much later, Enid wrote in the magazine, *The London Welshman,* of her decision to buy:

> Blaenwern suggests different things to different persons. To some it is a well-loved hymn tune; to others it is the headquarters of the Llanarth Stud and Farmers Ltd – and the home of the famous Welsh cob stallion, Llanarth Braint. And it is to Blaenwern that I, too, have come recently to live with my old friend and colleague, the cellist Pauline Taylor.

In many respects Enid, not only bought Blaenwern the family home, but she reconstructed the family estate as it was at the turn of the century. At a stroke she became a partner in the business and ensured that the Llanarth stud would continue but, unlike Barbara, her contribution was purely financial. She would not be a working partner. Pauline's attitude to money and finance was, to say the least, unorthodox. She once told Anne Wheatcroft, 'Money is there for me to use if nobody else is using it. Money is there to be shared. And it didn't seem to matter if it was other people's. It shouldn't lie idle.'

On reflection it was a source of conjecture and wonderment that the partnership with Barbara should have continued and endured so long, but then it was a meeting of opposites, both striving to make a vision work. Mercifully they lived apart. Pauline lived at Y Bwthyn and Barbara with her mother at Blaenwern. But when Pauline appeared, her presence dominated and Barbara took a back seat – in her own house. Betty Davies once said of her:

> There was always noise when Pauline was around – very entertaining, yes, – oh, yes, she liked people, loved them. She cherished old values and customs, but was always looking forward and ready to give new ideas a trial. Forthright, she spoke it to your back and she spoke it to your face, but never with malice.

Pauline, Enid and Barbara were friends, partners and unmarried. They were of a generation of women who, if they had dreamt of marriage in their early twenties, creating a home and bringing up children, their hopes were dashed by the shortage of men. These were the years following the end of the Great War in 1918, when three quarters of a million men had been killed. But, even before the war, there had been more women than men and the 1921 census revealed that women exceeded men by 1.75 million. Many had experienced loss and grief within their own families – fathers, brothers,

The Dorian Trio in the early days – Pauline Taylor (cello), Enid Lewis (piano) and Violet Palmer (violin).
Photo: Swaine, London

Two concert programmes from the 1930s for London audiences.

The Trio on the road in their trusted Jowett car.

Llanarth Braint showing his paces at the Royal Ascot Racecourse Horse Show, 1961.

Three palomino Welsh cob mares chosen to convey the stud's greetings for Christmas.

Blaenwern in its heyday on a sunny day.

An aerial view of the Blaenwern yard, stables, barns and the new shed in the foreground.

Llanarth Braint as an 18-year-old in August 1966; a Christmas card sent from Pauline to Barbara.

Llanarth Meredith ap Braint 8059.

Photo: Marston Photographs

Llanarth Meteor 3328.
Photo: Les Mayall

Llanarth Flying Comet 6835.

Sarah Hall and Barbara Saunders Davies (left) mounted on Welsh palomino mares rounding up some of the Welsh Black cattle for milking at Gilfachreda, Llanarth.

A tranquil scene. Mares and foals on pasture in the sunshine.

Photo: Carol Jones

Pauline Taylor preparing Llanarth Fortress for her last parade at the Royal Welsh Show (1971).

Llanarth Flying Comet, the Supreme Champion at the Horse of the Year
Show 1979. Len Bigley receives a silver tray from Princess Michael of Kent.
Photo: Findlay Davidson

Llanarth Flying Comet, Champion of Champions, modelled by Lorne Mckean at Royal Worcester Porcelain.

Name	No	Foaled	Height	Colour
Large.				
Llanarth Firefly	9114	1933	14.0½	Bay
Llanarth Firelight	9881 C	1950	13.2	Bright ches.
Llanarth Fortress	297. FS B	1943	13.2	Red roan
Llanarth Flying Saucer	1134 FS2 C	1951	13.2	ches.
Llanarth Dayspring	792 FS(3)391 PBASB	1947	13.1	Bright ches.
Llanarth Rhiannon	10499	1953	14.3	cream
Llanarth Nans	10501	1949	14.1	Bay brown
Llanarth Eluned	10676	1953	14.3	ches
Risca Pearl	10257	1953	14.3	Liv. ches.
Llanarth Madelon		1952	14.3	bay
Llanarth Beti Wyn	1482 FS.1	1947	14.2	grey
Llanarth Prydderi	G 313	1954	15	cream

Barbara Saunders Davies notes the year foaled, height and colour of each cob in the Llanarth Stud Book.

Llanarth Cerdin 3380.

Llanarth Flying Comet, Champion of Champions at home in Blaenwern
showing his ribbons and trophies, with Len Bigley and Pauline Taylor.

Photo: Idris Aeron, courtesy of Carole Knowles-Pfeiffer

Enid Lewis and Pauline Taylor about to begin a musical evening at Blaenwern.

Barbara Saunders Davies in reflective mood on her retirement, seated on St Davids Head, Pembrokeshire.

STALLIONS

AT STUD

SEASON 1958

AVAUNT G.S.B. Foaled 1950. 15h. 1. Bay Imported from U.S.A.

Sire : SUPER DUPER by BAHRAM.

Dam : GOLDEN HONEY by CHALLENDON by CHALLENGER II.

g.d. AOUDAIA by TOWN GUARD by HURRY ON.

A compact horse, full of quality with excellent bone, beautiful pony head and quiet temperament. Ideal for breeding show ponies. A winner of several races under Pony Turf Rules.

Fee : 8 Gns. (T.B. mares). **5 Gns.** (other mares)

LLANARTH BRAINT 1854. Welsh Cob, 14h. 3. Winner of fifteen first prizes, including Royal Welsh, Lampeter, Llandyssul, Newcastle Emlyn, Caernarvon, and Ponies of Britain Shows Has already established his reputation as a getter of Palominos and prize-winning stock. **Fee : 8 Gns.**

LLANARTH MARVEL 2644. Foaled 1956 Welsh Pony 13h. 2. Palomino of quality and substance. An outstanding mover. **Fee : 5 Gns.**

LLANARTH LUCIFER 2709. Foaled 1955, 13h. 2. Bright Chestnut. Grand pony. Strong riding type.
Fee : 5 Gns.

LLANARTH CRACKNEL 2710. Welsh Mountain Pony 12h. Palomino. Good colour and quality.
Fee : 5 Gns.

Mares met by appointment. Keep of mares can be arranged. Every care will be taken, but no responsibility for accident or disease.

The Misses Taylor & Saunders Davies
LLANARTH STUD

Blaenwern Llanarth, Cards.

Telephone : LLANARTH **235**

LLANARTH STUD
of Welsh Cobs and Ponies
including THE LARGEST STUD OF PALOMINOS IN GREAT BRITAIN

LLANARTH BRAINT

Supreme Champion Native Breeds, Ponies of Britain Show, Harrogate, 1958, where he won Overseas Visitors, Hanstead and Braishfield Cups. Consistent winner both riding and harness classes. Four of his daughters stood 2nd, 3rd, 4th and 5th to him in Welsh Cob Class at Harrogate. A great sire of Palominos. The Welsh Cob is Britain's finest ride and drive breed. Crossed with T.B. or half-breds, it produces splendid jumpers and event horses.

STALLIONS AT STUD:

WELSH COBS
LLANARTH BRAINT 1854, 14.3 hh., Fee 10 gns.
MENAI CEREDIG 2700, 14.2 hh., Dark Bay. Foaled 1955. Fine quality riding type with lovely head. **Fee 8 gns.**

WELSH PONIES
LLANARTH MARVEL 1854, 13.2 hh. Foaled 1956. Fee 5 gns.
LLANARTH SENTINEL 3057, 13.2 hh. Foaled 1957. Fee 5 gns.
Both these ponies are perfect Palomino colour and outstanding movers.

WELSH MOUNTAIN PONY
LLANARTH SPARKLE 2721. Bright Chestnut. Foaled 1957. A pony with excellent front—a free true mover. **Fee 5 gns.**

THOROUGHBRED
AVAUNT G.S.B., 15 hh. Brown. Imported from U.S.A. A grandson of Bahram. Beautiful pony head, compact, good bone and quiet temperament. **Fee 10 gns.**

STOCK FOR SALE   MARES KEPT BY ARRANGEMENT

THE MISSES TAYLOR AND SAUNDERS DAVIES
LLANARTH, CARDIGANSHIRE   (Tel: Llanarth 235)

R
B
&
B

Catalogue 2/6d.

LLANARTH STUD

CATALOGUE OF SALE OF

Welsh Cobs, Ponies of Cob Type
and Palominos

to be held at

Blaenwern Farm, Llanarth, Cardiganshire
on SATURDAY, 17th OCTOBER, 1964, at 2 p.m.

JOINT AUCTIONEERS :

RUSSELL, BALDWIN & BRIGHT
5 Church Street
Hay-on-Wye (Tel. 122)

J. J. MORRIS
Broyan House, Priory Street,
Cardigan (Tel. 2343)

The front of the catalogue for the first sale at Blaenwern, October 1964.

fiancés, uncles – and as Vera Brittain, who lost everyone close to her recorded, 'Only ambition held me to life.'

The press called them 'surplus women' but the terminology became far more contemptible as those women lost their youthfulness and were then labelled as being 'on the shelf', 'old maids' or 'sex-starved spinsters'. If they had brains they were 'blue stockings' – most men did not like clever or emancipated women.

In her book *Singled Out,* Virginia Nicholson chronicles the lives and attitudes of some of these women, and presents evidence to show the drawbacks of enforced singleness but also its potential advantages. There were those who managed to make new lives for themselves in a world that often ignored, blamed, ridiculed and even reviled the spinster. Anthony Bradshaw, Pauline's nephew said of his aunt when I asked about marriage, 'No, she didn't marry, but there were men of course. But she would dismiss them as not being satisfactory!' He went on, 'But one wondered what would have happened if she had met someone with the same farming interests, but I don't know really how to put it, she was too full of her own ideas and someone else might come in the way.' There is an element of Pauline's defiance in that observation, an opinion expressed when she was in her late middle age years, the course of her life already set. Yet looking back on the lives of Pauline and Enid, the reality of having to earn a living could have been grim, but they were independent women who took control of their lives. They fashioned careers in music and earned an adequate living. Barbara was somewhat cushioned from 'grim reality' with her background of wealth and privilege, she could have grown up expecting fulfilment through men, love, marriage and children, but she chose instead to study and later, to join her friends as they ventured to establish an unusual business which became successful gaining worldwide recognition. They became trailblazers and achievers, enterprising and independent but when Barbara decided to change direction to follow new pursuits in London, the next chapter in the story of Blaenwern took time to come to fruition.

Negotiations and the completion of the transaction for the sale of the estate to Enid was prolonged. In the meantime, animals needed to be fed and cared for, land to be tilled and fertilised, stock to be sold, an ongoing breeding programme and shows and sales to attend. Pauline found ad hoc help from friends, students and local farmers, but the first two years were difficult and hard. New living arrangements for Pauline and Enid were agreed as the transaction was completed. Pauline would move from Y Bwthyn into the Blaenwern house with her cats and Sianco, the loyal Cardiganshire long-tailed corgi. Enid would maintain her cottage, Rhydfawr, but when she fully retired from the Guildhall, she would join Pauline and live at Blaenwern. Her article written for the *London Welshman* in 1964 sums up her sense of family and of her roots:

> My forebears were farmers and my grandfather a well-known breeder of Welsh cobs – maybe there is something in my blood which has drawn me from my native Glamorgan to Cardiganshire and to Blaenwern, my maternal family home. I know little about farming but I am trying to learn and, strange though it may seem, I don't feel that music is too far away. After all, my mother was a musician reared at Blaenwern.

But practicalities were always uppermost in her mind:

> Our main aim and very great concern is to try to rehabilitate our Welsh native breeds which always prove their worth in quality and hardiness. But financial worries are never far away from the ordinary person these days and in farming we certainly have them. However, I hope that we may be able to weather the initial money storms and make a 'go' of what is a unique Welsh enterprise.

Part of the agreement when Barbara left Blaenwern house was

to leave much of the furniture which she had inherited from Pentre behind in its place in the house. The flat in London she was moving into was far too small to accommodate oak corner cupboards, chests, tables and chairs and Pauline was able to add her own pieces of Welsh oak furniture – not that she was an interested homemaker or housekeeper. The general consensus about the house in those years was that it was welcoming but chaotic, with shelves, tables, chairs, any flat surface weighed down with books piled high, papers and magazines and a thickness of dust, with corgis and cats almost 'in charge'. Enid was the complete opposite. She had style and taste; her flat at Fitzroy Park in London was a haven of warmth and elegance, as was Rhydfawr, her Cardiganshire cottage. When she eventually moved to Blaenwern, she soon stamped her influence on her south-facing sitting room overlooking the lawns. Enid was a homemaker.

She gave Pauline her full backing but it was not until five years later, in 1969, when she finally retired as a revered teacher at the Guildhall School of Music that she could give Blaenwern her commitment and Pauline her full support. She began coaching and teaching and she found huge satisfaction when she discovered a twelve-year-old girl, Gwawr Owen, who went on at Enid's prompting to further studies at the Yehudi Menuhin school. The final mortgage payment on the Blaenwern estate was paid from the sale of her London flat and as a former colleague at the Guildhall said of her at the time of her retirement, 'A near lifetime of hard work has gone to the recovery of the estate, and she gets great joy from the spectacular and sustained success of the farm. Enid is enjoying every minute of her 'wildly active retirement'.'

Somehow or other Pauline had survived the first two years of farming on her own by sheer guts and tenacity and 1964 proved to be another turning point – Len Bigley arrived at Blaenwern.

CHAPTER EIGHT

I T HAD ALWAYS BEEN Blaenwern policy to encourage young people into farming and when Pauline received a letter from a sixteen-year-old student from Bishop's Castle, Shropshire asking if there could be a holiday job for him, she sent 'a lovely reply', ending with the invitation, 'Come for a month and see how you get on.'

Len Bigley was intending to study agriculture, 'It was 1964. Pauline was desperately looking for help with the stud which had grown steadily since the 1940s. I went for the summer, thought things over and, after a month or two, I never returned home.' When he arrived at Blaenwern, 'There was a very nice lady and her husband from Ireland. She was a sort of house keeper and she cooked, but unfortunately she only stayed six months. So for that time I was quite well fed.' The house was chaotic and it remained so, people came, people stayed and, 'There were cats, at that time thirteen cats – and ever since then I've never allowed a cat in the house – and dogs, three Cardiganshire corgis, those with a long tail.' Len remembers that first day with a certain amount of pain:

> Carole Pfeiffer, a former student was helping out and she showed me the stallions, and I know I didn't put the one bolt in the stable door properly, and the stallion came out of his stable and made straight across the yard to fight with another stallion through the top door. Never having dealt with stallions, I raced across and tried to separate them. Never do that because, rather like dogs when they're fighting, don't try to separate them. You have to be very careful. When I got across the yard, the stallion kicked out and caught both my knees. I caught the stallion but I was feeling very, very sorry for myself. They were two good blows. We put the stallion in his stable and Carole said, 'Come on, I want to go to the next farm.' She started running through the woods and I ran after her, my two legs in agony. That was my first day at Blaenwern.

Len had learnt a hard lesson but he had a natural empathy with animals especially horses and he had a golden opportunity to show his mettle. He found Pauline demanding, inspirational and kindly. She could see in him a natural aptitude for handling and training horses. He was strong and he had patience – his quiet demeanour and firm instructions inspired confidence and even the most lively and excitable stallions were soon disciplined through horsemanship, kindness and understanding. 'If you get through the first week or two without shouting you will succeed. You have to understand their temperaments and personalities, it is something you learn and I enjoyed the job.'

Despite all her tribulations following Barbara's departure Pauline never lost sight of her dream of holding an auction sale for Welsh cobs and cobs of pony type and when Len made his decision to stay, she soon set the date for the first Llanarth sale, Saturday 17th October 1964 at Blaenwern, which, according to the *Old Farmer's Almanac* heralds a spell of fine weather and is so called 'St Luke's little summer day'. Since the sale was to be held in the 'open air', she needed all the help the saints could muster. Pauline had determined that a sale of pedigree Welsh breeds should be held in Wales and a sale of pedigree Welsh cobs, Sections C and D should be within the cob's natural home, Cardiganshire. During the past two or three years, the number of horses kept at the stud had almost doubled, far too many for the size and nature of the farm, but the priorities for the sale set by Pauline were twofold. Firstly, to sell the Blaenwern horses and secondly, to give local breeders the opportunity to sell their animals. The event was a gamble and when I asked Len Bigley how they coped with organising such an occasion, 'It was a good idea. It made us tidy the place up. It made us put an order where there hadn't been one since Barbara had left.'

Tidying up meant brushing and cleaning and putting an odd pot of paint on farm buildings and stables, and also clearing the garden entrance to the house. Len recalls the state of everything, 'For instance the rhododendrons which had taken over the garden, the drive and lawns, the grass hadn't seen the light of day for goodness knows how long. It took weeks to cut it right back and to clear it but it was brilliant, it all looked welcoming.' The two years when Pauline had struggled to keep going on her own had taken its toll, but with Len's enthusiasm they began organising the sale and promoting it to local breeders. She went to a local auctioneer from Cardigan and, although they agreed to conduct the auction, they found such an event, a pedigree cob sale, quite daunting.

That first sale attracted 63 entries. William Lloyd remembers his father taking a young foal to one of the first Blaenwern sales, 'Miss Taylor spent some time persuading him to take part, and he was one of the first of the old breeders to recognise Pauline and Barbara as serious breeders. But I have to admit that that first sale posed more questions than answers yet somehow there was an air of expectancy and optimism around.'

The catalogue was splendidly produced, showing 63 for sale, 35 from the Llanarth stud and 28 from other breeders. Pauline wrote an explanatory foreword:

> Llanarth stud is the largest stud of Welsh cobs and ponies of cob type in existence and was founded some twenty years ago. The aim was to breed a type which, by losing nothing of the true native quality, substance and stamina of the old fashioned cob, would also have real riding quality and, above all, free forward action from the shoulder.

She also included a potted history of each cob, noting sire and dam, plus one or two colourful observations such as, 'Llanarth Chanel: A filly likely to breed palominos. An excellent mover. Quiet and halter broken. Llanarth Hefyn: Bay gelding. Grand shoulder and action. Quiet and halter broken. Training not carried further due to kick on nose some months ago. Werndriw Mattie: Dark chestnut filly. True Cardiganshire cob. Very active.'

October 17th dawned into a dark and dismal day but the sale proved to be a significant event. It brought the community together who helped with last-minute arrangements and welcomed buyers from far and wide. The event enticed Barbara to return to Blaenwern, answering a generous invitation from Pauline. There was no animosity and she received a warm welcome from everyone, which made her realise that her original thoughts of staging a sale had been incorrect.

The top price paid was 300 guineas for a palomino filly, Llanarth Sissel, bought by breeders from Yorkshire, and a champion stallion, Gwynau Boy, sold for 200 guineas. The sale total for the 36 animals sold was over £3,700 with an average of £104 per sale. Bidding for 15 cobs did not reach the asking price and were withdrawn and, in that sense, it was disappointing but Pauline, as ever, put her faith on the line and proclaimed this would be an annual event. Every year, the numbers increased and prices went up and up to buyers from England and overseas. A decade later, the number of horses for sale had trebled and in 1973, nine years following the first sale, the financial total was £58,537 and the 1979 total reached £82,010. Records had been broken. Pauline's foresight was proved correct; year after year the cob's reputation was gaining recognition, the breed's attributes was winning approbation and the breeding policy respected and admired. The annual autumn sale had become *the* cob event of the year in cob country.

Blaenwern was a meeting place, with Pauline, the central figure, talking, gossiping, arguing and entertaining, sharing a drop of the 'hard stuff' and indulging in her latest habit, inhaling snuff. Many of the old boys rejected that particular way of receiving their daily input of nicotine – they preferred pipe smoking, cigarettes, cigars and few old-timers persisted with their habit of chewing tobacco. Enid would also always return from London for those early sale days and she would hold court in the kitchen and preside over the hospitality in the house, but she drew the line at inhaling snuff.

Barbara never once returned after that first sale and that was a particular sadness for Pauline, but her success more than compensated.

But Anne Wheatcroft remembered the high days, 'Everyone would come and pile into the house, into the kitchen, sitting room, dining room, drawing room, all of them well-known Cardiganshire breeders and farmers, drinking out of their skins. We had some great fun and Pauline was at the centre of it.' She paused, remembering, wondering whether she should continue, looking out of the window, a broad smile on her face picturing the scene at Blaenwern – the Aga going full pelt, the old oak tables and shelves packed with magazines, books and papers and the dresser, festooned with colourful ribbons and rosettes plus silver cups and medals won at prestigious events, the noise of laughter, storytelling and argument filling the night air:

> And then we went one evening to a pub, not far from Cross Inn; we all went in there to have a musical evening. All these chaps arrived in their ponies and trap, tied them outside, all these old boys, lecherous some of them, trying to sit next to the nearest lady in order to run his hand up your thigh. Pauline would stand up and retreat. But what laughter, raucous laughter and risqué stories.

Enid's financial support and the successful cob sale lifted Pauline's spirits and brought a fresh impetus to her farming activities. The event was also helping the local economy. Hotels and B&Bs benefitted from buyers who stayed in the area; it generated a close community spirit, with farmers and sons helping with traffic, parking and stabling, and wives and daughters serving endless cups of tea and sandwiches.

Len's decision to stay at Blaenwern and to become a

full-time groom eased the pressure on Pauline's mental and physical strength. The 1964 sale had reversed that feeling of dejection, and Enid wrote of her excitement at the duality of her personal life, 'I think I have the best of both worlds – enjoyable music-making in the holidays without the formalities of the concert platform and the worry about the financial side of things.' The farm's general economy was in a fairly parlous state because of a lack of investment for land drainage, new buildings and general upkeep of the old. The stud was paying its way and the tenant farmers at Penlon and Frongoch were making a reasonable living, but sales of cobs, although doing well, could not by any means uphold the total investment needs. When Barbara left, Pauline spent a high proportion of her farm income on labour costs and whenever she borrowed money, the banks, as was their custom in their dealings with women, put pressure on them with high interest rates on their overdrafts. It had not been easy to be a woman in business on her own. But Enid's obvious delight in her ability to re-create the estate around Blaenwern almost as it was in her grandfather's era brought contentment and with Len's arrival there was also hope.

In two years Enid would be seventy-five and Pauline seventy, both defying advancing years and looking forward with excitement. Pauline often revitalised her weariness and her energy levels by giving way to short cat naps during the day wherever she happened to be: in mid-conversation at dinners or gossiping at informal gatherings, at meetings, shows and sales. The most dangerous occasions would occur when she was driving the car seemingly 'napping' without warning. She had always been a somewhat adventurous, if eccentric and erratic driver, even in her younger days when she first began farming with Barbara. They had bought an old black Jowett, some say the first of that model to arrive in Cardiganshire. Then came a well-worn canvas topped trailer to carry their stallions around the highways and byways of

Cardiganshire during 'the season'. Their journeys were rarely without incident – the weight of two Welsh cob stallions in a 'none too sturdy trailer' caused the Jowett to sway a little (a model which had the same type of engine for twenty-eight years and had gained the reputation of always 'getting there'). It needed a driver of nerve to drive through the narrow Cardiganshire lanes and Pauline had buckets-full of nerve. But, an unusual streak of meanness would attack her in the car, especially when she approached those long undulating Cambrian hills of mid-Wales.

She would put the gear in neutral, placing her right foot on the brake pedal as she turned the engine off to go down a long hill. In the early years freewheeling in the Jowett was Pauline's way of saving petrol. As Len recalls, 'These were hair-raising experiences, especially racing the downhill gradient in neutral and the sharp corners of Plynlimon hill.' Somehow she accomplished those journeys without absolute disaster, but there were accidents. One in particular occurred in 1965, when she was on her way to the dentist and she crashed and was seriously hurt. A smashed hip kept her in Aberystwyth hospital for ten weeks. Frustrated and dependent on others, she recovered to be active again except for the use of a walking stick. By this time, she had changed the Jowett for a Ford Anglia and this was later superseded by a Renault, but the erratic driving continued. Her sister Alice was equally capricious behind the driving wheel of her little Austin van. Thankfully, she rarely drove in the summer because the lanes around Gilfachreda were too busy with visitor traffic, but she would join Enid and Pauline once a week for Sunday lunch. She often forgot to engage the hand brake when she parked on the sloping car parks of west Wales – a habit which often caused mayhem in towns and villages. One such incident occurred on a Sunday when she arrived for lunch at the Blaenwern sloping yard. She got out and went to get something from the back of the van and, as she opened the door, the van slowly moved backwards knocking Alice to the ground. Len streaked across the yard as the van was gathering momentum, flung his weight behind the van to stop the vehicle from running right over her down the slope. 'I held it but I couldn't push it up the slope but thankfully we managed to pull her out before any damage was done.'

If that was a narrow squeak, lady luck shone brightly one evening when Pauline and Enid, Betty Davies and Len were approaching Aberystwyth on their way to a concert. Pauline was driving. Although a fairly tall woman, she always seemed to sink lower and lower in the driver's seat as she drove. The Ford Anglia would often appear driverless to other cars – the only sign of life being a few wisps of grey hair to be seen between the spokes of the driving wheel. No-one really can account what exactly caused this incident. It was dusk, they had come down the hill to the village of Llanfarian, foot on the break, turned a corner to approach a narrow overhead railway bridge when an articulated lorry came towards them. Too late to avoid, both drivers attempted to squeeze through at the same time. The Ford Anglia lost the battle; it went straight underneath the articulated trailer. It peeled the top of the car off completely. Len vividly remembers the shock and the presence of mind, 'Pauline was so low down in the seat it didn't touch her. Enid was tiny anyway – it didn't affect her. Betty and I simply ducked. Thankfully we were all alright.' And he summed up Pauline's exploits behind the wheel thus, 'Flying down hills in neutral was very brave or completely mad. She could see madness in other people, but I'm not sure she could see it in herself. I think she probably thought she was the sanest person on earth.'

CHAPTER NINE

A S THE 1960S DREW to a close, Pauline's farming worries had eased considerably and the Llanarth cob breeding programme was producing a long line of champions. Len was not only an accomplished handler but he was becoming an astute judge of the cob breed. Llanarth Braint, the grand champion born in 1948 (by coincidence the same year as Len Bigley) remained supreme. Twenty years later (1968) he sired Llanarth Flying Comet – another supreme champion in the making, followed a year later by Llanarth Meredith ap Braint, a much sought-after stallion who sired Llanarth Brummel, another winner unbeaten under saddle and owned by Anne Wheatcroft. Braint retired from the show ring after winning the championship at the 1967 Royal Agricultural Society of England Show and Flying Comet was a splendid successor. He was never beaten at the Royal Welsh Show and qualified four times for the Horse of the Year Show at Wembley where he was champion in 1975 and supreme champion in 1979 and 1980.

Len proved an expert handler who turned out champions in top form year by year. He would run a cob in hand in one class, followed perhaps by a quick change of clothing to show another under saddle and then don a smart topper to compete in the driving class. 'I just enjoyed the job I think. I like working with animals. It's something you learn. When you have a leading rein in a show, you try and get the same trotting pace as your horse. You need to put in time at home to practise all the skills.'

And as the 1970s dawned, an enormous purpose-built building was erected which enabled Len to stable, store feed, train and show his horses in all weathers under cover. The building was divided into sections under one roof. There were five boxes to house stallions lined on one side and on the other an adequate space for hay and feed plus storage for halters, bridles, cleaning materials, brushes and combs. In the centre there was a large circular exercising area which proved a boon for schooling and also for showing stallions, mares and foals for visiting buyers. This major investment was opened in 1971, the year that Enid showed her faith in the future of the Blaenwern estate now that she had fully retired from the Guildhall School of Music and Drama. She was eighty years old and ready to face new challenges. At the Guildhall, James Gibb, a former colleague and renowned teacher and pianist gave his fulsome tribute in the Guildhall's *Review* with these words:

> No one, they say is indispensable. But surely there are some who are irreplaceable, and, just as surely, Enid Lewis is one of these. No attempt was made to hand over her widely varied teaching responsibilities to a single successor when she retired at the end of last summer term, and indeed it would have been a most daunting task to find someone with comparable diversity of musical experience matched to such qualities as Enid possesses. There was something unmistakably and endearingly Welsh about the intense personal concern she showed for her students and the need to make true personal contact. No teacher got to know more about her individual pupils, and never by inquisitive probing or solicited confidences, but from the mutual trust and respect she always inspired... Enid is in fine fettle, coping with a whirligig of activity on her estate in Llanarth, Cardiganshire, the estate which belonged to her grandmother. A near-lifetime of hard work has gone to the recovery of the estate and to the final payment of the mortgage, now complete. Although Enid professes not to know one end of a horse from the other, she gets great joy from the spectacular and sustained successes of the farm. No doubt Blaenwern will continue to have many visits from all of us who have loved and admired her all these years.

Enid moved into the old house, her grand piano taking pride of place in a large room of her elegant furnishings and paintings overlooking the lawns and the garden. Although she thought she had finished teaching she could not resist taking under her wing the gifted local twelve-year-old girl, Gwawr Owen:

> I remember her well that first day. She was dressed in a navy blue dress with a little coatee jacket and she had the most wonderful chunky necklace of different blue hues. She was always very well-dressed, always dark colours. So different from Miss Taylor who was a working farmer, out and about, noisy and not really caring how she looked.

At that time Gwawr was very competitive, having been brought up in the Welsh Eisteddfodic tradition:

> Miss Lewis was tremendously supportive and wherever and whenever I went, the local, the National or the Urdd National Eisteddfod she would come with us. It didn't matter how many were competing in the prelims, she would sit from beginning to the end, that is, as long as my parents could get her a cup of tea and a Welsh cake, she was happy. She was over 80, quite frail, couldn't walk very far, but she was genuinely interested in my career.

Gwawr had been mainly taught by local teachers until Enid Lewis took over, and it was at a concert in the Wigmore Hall and a chance meeting with one of the teachers at the Yehudi Menuhin School, Mrs Chappelle, that Gwawr was encouraged to apply. She was accepted to spend five demanding, at times difficult, years. She was older than the majority of students and it was her first experience of being away from home. Her recollection of Enid as mentor and a teacher was warm, kindly and respectful:

> She was inspirational. I wasn't one for practising but she taught me so much, how to work round problems, for instance, a run of notes – the fingering, and if I didn't get it right she'd say, do your fingering on the piano lid, the wood without the reverberation will show you exactly what's right and what's wrong.

The musical evenings at Blaenwern continued in old age. Enid and Pauline would often play together to relax after a day's work or listen to Radio 3. They attended concerts at the University College in Aberystwyth and occasionally Len would drive them to London for a particular concert. They would spend the night at their club and Len would be at a nearby B&B. Music remained an integral and important part of their lives, or, as Anne Wheatcroft expressed, 'Two interests, music and horses – they go together like art and horses. Its form and movement, there's beauty in both.'

The year 1971 was notable for Pauline. She was elected President of the Welsh Cob and Pony Society and her message to members, as she began her year of office was:

Happily the occasion coincides with the rising popularity of the breed as the recent sale at Llanarth amply proved. My first love, at the tender age of eight, was a dun Welsh cob and later on my dream of breeding them was to be fulfilled. Now after thirty years of breeding may I, on reflection, pass on what I think I have learned over the years. If we aim to breed the best, then showing must take second place to our breeding, and any material reward for our work that may justifiably be expected must be sought last. If we keep our priorities in the correct order, the best will come almost unsought.

Pauline's enthusiasm seemed limitless and infectious, her energy at the age of seventy-five was a source of amazement as she attended shows and sales and meeting after meeting, occasionally cat napping if discussions became tedious but always waking when matters became interesting, and even at times able to recall what had been said while she slept! She reflected to a journalist once that she was following her childhood dream, 'The whole thing is like a miracle. I never thought I would do it.' Wynne Davies who chronicled the society's history in his admirable tome, *One Hundred Glorious Years*, opens the chapter on the activities of 1971 with these words:

1971 will be remembered as the cornerstone year for the Welsh cob. It is difficult to pinpoint any one cause as being responsible for this unprecedented surge of interest and demand. There were many contributory factors such as the Presidency of Miss Pauline Taylor of the Llanarth stud.

Highlights of that year included an invitation to cob breeders to assemble cobs for sale and display them before the King and Queen of Nepal and the Gurkha horse master, who wished to purchase two. They were well-satisfied with their purchases – a mare, Aber Bess and a yearling filly, Trefaes Trixie. Llanarth Brummel was invited to Buckingham Palace for the display of British breeds in honour of the King of Afghanistan, but without doubt, the main event was the display of cobs, sections C and D, at the Royal Welsh Show in July.

Since its inception in 1904, classes for horses had been at the heart of the show's events and the interest has grown year by year, and today at the modern show it is as intense as it ever was. The grandstand and the grassy banks around the main ring on the third day of judging are always filled to capacity, especially so during the afternoon when Welsh cob stallions and mares, section D – often over twenty in number are judged for the championship. It is an event when handlers such as Len bring out the best in their cobs as they run in partnership and symmetry with the action of their animals, the cheers, whistles and applause from supporters reverberating around the arena.

At a special display in 1971, 52 animals were selected to parade and show the versatility and temperament of the Welsh cob including, of course, its beauty and strength. Cobs in harness, ridden, shepherding, carrying disabled riders, dressage, side saddle riders in Edwardian dresses were led into the main show ring by Miss Pauline Taylor. Never one to miss a promotional trick, she rode thirty-year-old Llanarth Fortress escorted by three ridden stallions,

Meteor, Rhodri and Rhys, each one descended from Fortress. The commentator, Wynne Davies made the point that the combined age of rider and Fortress exceeded a century and this provoked a huge round of warm applause. The success of the display prompted a surge of enthusiasm for an invitation to display cobs and ponies at the Essen Equitana in Germany in April the following year and Pauline, with customary zeal, took a group of cobs to show at the event. One German buyer, who had never heard of the breed, bought a Section C pony, Dwyfor Peter Pan, a grandson of Llanarth Cerdin. He had been the top priced C pony at the Llanarth sale a few months earlier. All these events increased interest in the breed and in the important export trade. Pauline was always an eager participant at the Royal Welsh Show, showing the cob's versatility as the 'working cob' and the discipline and intricacy of dressage displays. Occasionally she would assist Wynne Davies in the commentary box and once or twice the microphone which she thought was switched off would pick up her blunt observations, 'I wish that stupid man would stop and get out of the ring, he's gone on long enough.' Straight talking for one exhibitioner and laughter from the crowd. That sense of community and competition, the meeting of town and country, plus the sight of animals in all sections presented with pride gives the show its unique quality, pride, warmth and happiness.

A huge crowd came to Blaenwern on a glorious autumn day that year to support the Llanarth sale. There were 191 cobs in all, the section C horses sold for an average of £120 and section D for £187 with the top price paid for a filly foal reaching 500 guineas. Llanarth offered 29 cobs for sale and sold 23. They were well-pleased with the prices except when the bidding for Llanarth Jenny Jones, a stunning filly foal reached 470 guineas. At that point, Pauline withdrew the foal in the belief that she would fetch at least double that amount the following year. She was right. Jenny Jones was sold for

1,000 guineas in 1972, the first filly to reach such a figure. But the sale ended on a sour note. Three top lots were paid for with a stolen cheque and they left the sale yard in a stolen horsebox, but all ended well – the perpetrators were caught and sent to prison.

Within months of her appearance at the Royal Welsh Show, Fortress died and it prompted this tribute from Pauline in the society's journal:

> She left us on 16th November in her thirtieth year as quietly as she would have wished, while grazing with other mares – no fuss, no nursing and no trouble. Never, as I can remember did she need a vet – healthy, full of zest to the end. Always walking out with a brisk purposeful stride and at the slightest hint, break into a gay trot or an easy, smooth canter. In days long ago at harvest time, she would canter me swiftly with a basketful of crockery on my arm for miles across fields without the rattle of a tea cup. How well I remember the lucky day we saw her and bought her twenty-seven years ago. Her mating with Llanarth Prince Roland proved to be one of those critical moments in the evolution of the Llanarth stud. From the union came a chestnut filly, Llanarth Rocket. The first filly foal in 1951. I took Fortress to the Royal Welsh cob display. For her, with her progeny around her, Meteor, Rhodri and Rhys, it was a fitting climax to the life of a wonderful Welsh pony.

For a decade and more Pauline had been involved in a charity called the Welsh Society for the Humane Disposal of Surplus Ponies. She joined forces with another breeder, Sheila Richards, because both had become increasingly concerned at the sight of young foals, just weaned from their mothers, being trailed in a weakened state from market to market in an effort to find buyers, often ending up in ignorant hands for a few pounds. They received official backing from the Welsh Pony and Cob Society to buy some of these foals at markets and auctions and to rear them. Following extensive research, they found a kindly Bristol horse abattoir owner to cooperate with them in order to humanely dispose of the weakest. For instance in 1976, the society bought and cared for a total of 188 foals, 17 yearlings and 5 adults. Today it has become a recognised and well-organised charity supported by the main pony societies. Sheila Richards, who had run the organisation since its inception was awarded the British Horse Society Equine Welfare Award but she died before the Welsh Pony and Cob Society in 1990 could award her their Life Vice-Presidency. She received it posthumously. Pauline was awarded a similar honour fifteen years earlier, marking her notable contribution and presidency. Barbara's contribution to the society was not forgotten as she had been an active member since 1943, and for her long service as a judge, she was elected Honorary Life Vice-President. Pauline had also been President of the Welsh Black Cattle Society and following her year of office, they elected her Honorary Life President, as a mark of appreciation. Although there was no obvious sign that age was causing her to relax and to ease her activities, she and Enid had begun discussing and planning for the inevitable.

Time was running out and the constant theme of their discussions was the question of retirement and who could inherit the estate. They deliberated as to whom in their families was equipped financially and had the interest and skill to manage the stud and the estate and to keep both as an entity. One name came to mind, Professor Anthony Bradshaw, Pauline's nephew, who had over the years often visited Blaenwern and shown an interest in farming and their activities. He was the youngest son of Pauline's sister, Mary Taylor, an archaeologist, married to Professor Harold Bradshaw, a renowned architect based in Liverpool.

Tony was a Cambridge graduate in Botany, gained his PhD at Aberystwyth which led to a lectureship at University College, Bangor and, to my great surprise, when I began researching into the lives of Pauline, Enid and Barbara, I discovered it was the same Tony Bradshaw whose lectures in Agricultural

Botany I had attended at Bangor during my undergraduate days. He became the revered Professor of Botany at Liverpool University, a pioneer of restoration ecology for his fundamental research into restoring vegetation in polluted soils and functioning ecosystems to derelict land. Sadly, he died in August 2008, but he is remembered as an inspirational teacher of clarity and enthusiasm and I was lucky enough to spend a morning with him at his home in Liverpool talking about his memories of many visits to Blaenwern. He showed me letters, books and papers relating to Pauline's family and The Dorian Trio.

He had visited them many times when he was working at Bangor and continued to keep in touch. Pauline and Enid invited him to Blaenwern to discuss their plans, 'Towards the end they weren't certain what they could do with it. I suppose if I had shown any interest, I could have had the farm and probably inherited the land and everything. It was the last thing I wanted to do.' It was the early 1970s; Tony Bradshaw was at the height of his powers with research and teaching:

> There were two reasons – one purely agricultural. A lot of the land was poor. Waun – it could have been improved of course, but it was not my idea of what farming was about. It was old-fashioned and needed a huge investment. Secondly, if I had the farm to run, I knew Pauline would never leave you alone. It would never have worked, we would have fallen out in no time.

Pauline and Enid had already made Len Bigley a partner in the business for his outstanding work with the stud. His contribution had been immense and they had come to rely more and more on him, but they were anxious to keep the estate intact and to preserve the core of the stud and its breeding programme. Enid, in particular, did not want the estate split up after a lifetime of work. They turned to the Welsh Pony and Cob Society. They offered to gift the stud and farms to the society as a basis to perpetuate and develop the breed, but the society was not able to accept the offer as it lacked the finance needed to ensure the upkeep of the farms and, as a society, it could not get involved in nurturing one particular stud.

Next they approached the University College at Aberystwyth. Over the years Pauline had developed strong connections with the institution not only through music-making but also as a member of the University Council. She often met influential leaders of its affairs. The College had a strong agricultural department and Pauline passionately felt that Blaenwern could become a centre for studies and research into the special qualities and development of native Welsh breeds and Enid believed that the College would be the custodian of a typical Welsh farm estate. They met the then Vice-Chancellor, Principal Goronwy Daniel and the President, Sir Ben Bowen Thomas, and after a fruitful and enthusiastic discussion, both men believed the scheme could be viable. There would be further detailed discussions into the legalities of such a gift and how the scheme would operate but they agreed that Pauline, Enid and Len would continue living on the farm. As they ended their initial discussion, they both felt their concerns for the future had eased.

CHAPTER TEN

PAULINE AND ENID BELIEVED that the University College could create a basis for a national stud for Wales, the first in the UK. It was another dream, another vision. Enid was satisfied that her investment, which had taken most of her life savings, would be her bequest to the nation. It would ensure continuity and the estate would remain as an entity as it was in her grandfather's time. Their enthusiasm knew no bounds. They had placed their trust in the two knights, Sir Goronwy Daniel and Sir Ben Bowen Thomas, two men with impeccable reputations as public servants who 'got things done'.

Their goal was to complete negotiations in readiness for the Annual General Meeting of the Welsh Pony and Cob Society at Llandrindod Wells in April 1974. At that meeting, it was announced that the University College of Wales, Aberystwyth had accepted the donation from the joint owners, Enid Lewis and Pauline Taylor – 385 acres of Blaenwern estate land and the prestigious Llanarth stud. Sir Goronwy Daniel attended the meeting and confirmed that the University would do everything possible to foster the breeding and development of cobs at Llanarth. He referred to the agreement as, 'a major event'.

And that year, 1974, the stud achieved a major success when Llanarth Flying Comet won the premier award, the George, Prince of Wales Cup at the Royal Welsh Show. It seemed the perfect accolade with which to end their ownership of Blaenwern and the Llanarth stud. The cup had a long history. It was donated by HRH George, The Prince of Wales in 1908 to encourage the breeding of quality Welsh cobs in order to keep the correct gene pool and bloodline from other influences and crossbreeding. At the time of writing, 2009, the cup has been awarded 101 times, and over 50 of those champions have been bred at Cardiganshire studs. The quality of Llanarth Flying Comet reached its peak on three more occasions winning the cup in 1976, 1977 and 1978.

The future was assured, the University College had accepted the gift and the relief for Pauline and Enid restored their confidence and optimism. They would continue to live in the farmhouse, allowing the college to have a room to use as an office and they would bear the cost of heat, light, and telephone. Enid was happy with the arrangements, she wrote, 'As you know I wish it to be run as a Welsh farm with Welsh Black cattle, Welsh cobs and Welsh pigs. I think the University College knows this and I also think they would be happy to carry it on as it is now being run.'

The agreement clearly stated that the land was gifted by Enid Lewis but the animals, which included the cobs and the other Welsh breeds were bought by the University College for £10,000, a deal which cleared Blaenwern's outstanding debts. Reports in the press noted that the Llanarth stud was an integral part of the gift. This was misleading and incorrect. Legally the stud and other animals were not gifted, they were bought by the University College.

During the negotiations Pauline and Enid had also failed to recognise the true significance of one crucial clause included by the college legal team, 'that the farm could be used for the general purposes of the college'. It was the legal interpretation and implication of this clause which was to cause deep divisions, anxiety and heartbreak in the next few years. Given the benefit of hindsight, they realised there had been one serious omission. Enid had not insisted on a defined clause to safeguard the estate as an entity. She did not want the estate to be sold piecemeal, although it was without any sense of foreboding that she wrote to her solicitor on such matters:

> You will see that I have crossed out my consent to the letting of any part of the property not even at any particular time. If this was left

in, the University College might want to let later. Naturally I don't
anticipate anything like that, but I don't want to risk that happening
in the distant future.

They realised the major weaknesses of the agreement, in
addition to the lack of clarity and understanding, two or
three years later when Pauline and Enid came to interpret
a particular clause. They had been unwise to engage the
solicitor to act on their behalf when he was already acting for
the University College. It was to prove disastrous but it also
proved how trusting they had been of the College officials,
and surprisingly how naïve.

A charitable trust was established and when the agreement
was signed and sealed in 1976 by both parties, the future
seemed assured. A public announcement was issued and
Pauline Taylor was to make this statement to a television
reporter:

> I do think it's essential to remember that these breeds of ours have
> a long history. In Wales we haven't had the wealthy landowners who
> have been behind say the Scottish or English breeds and that is what
> is lacking here, and I wanted to get a body of some prestige such as
> the University College to take them up and see that they had fair
> play.

Every discussion and negotiation leading to the signed
agreement was held against a background of financial crisis
and an economic downturn in the country. Inflation in 1975
peaked at 25 per cent, to rise two years later to 27 per cent and
this, coupled with rising prices, high taxes, the declining value
of sterling and inflationary pay claims, rising unemployment
and widespread strikes in the public sector added to the
gloom and led in 1978/79 to 'the winter of discontent', a
time when rubbish not collected piled up on streets and the
dead remained unburied for weeks. The decade ended with a
general election sweeping Margaret Thatcher into power and
eighteen years of Conservative government.

There had been very little investment in the land and buildings at Blaenwern during the years leading up to the agreement, although income from the Llanarth cob sales had almost doubled between the first two years of the decade, but so too had costs of labour, feed, transport to shows and events. A film shot in the winter of 1976 by a member of the University's Agriculture Department and edited without commentary has been kept at the Film and Sound Archive Department in the National Library of Wales, Aberystwyth. Entitled *Blaenwern before Improvements*, the first half, shot on a grey mid-winter day, was not a celebration of a notable gift, but a bleak and stark picture of an upland farm where other holdings on the estate had been allowed to depreciate and languish. The film showed grey-stone buildings in serious disrepair, hedges uncut, stony and marshy fields, old gates hanging off their hinges, and as the camera panned around the landscape, a few fields seemed to be reverting to scrubland of brambles, Molinia grass and reeds. It seemed unnatural. There were no signs of cattle, sheep or horses except for one solitary palomino colt grazing quietly. Other holdings of the estate – Penlon, Dolau Llethi, Gofynnach Fawr, Soar, Frongoch are shown in a state of disuse, deterioration and seemingly redundant. Even Enid's first home, Rhydfawr, the pretty whitewashed longhouse by the stream, has an air of resignation about it. There is no sign either of Enid's major investment in the large new building at Blaenwern for stabling stallions and schooling young horses until the second half of the film, post the legalised agreement and the April takeover. The scene then changes to spring/summer colour. It shows drainage digging underway in a marshy field, the first sign of irrigation and the formation of a pond, two or three fields appear to have had a dose of lime and the cobs and cattle have returned to graze spring grass. There is a feeling of springtime regeneration in the wide angled vistas, but during that first year Enid and Pauline had come to realise that imperceptibly life for them had subtly changed. Others controlled estate affairs and Pauline especially found her status diminished, although she was referred to as a 'consultant', her involvement with the cobs was restricted. Len Bigley, as a partner in the business before the agreement took effect, remained as the stud manager answerable to the new farm manager, Jim Lees, who, unlike a few of his colleagues, knew a thing or two about horses as his wife ran a successful riding school as he coped with the stud's situation with growing anxiousness. Pauline had never been afraid to voice her opinion but Jim Lees, sympathetic to her views, was however listening to other masters who were intent on reducing the number of cobs at the stud, improving the land, increasing production and making the farm financially viable.

Pauline, Enid and Len continued to live in the farmhouse but ownership of the house, land and animals had passed to UCW Aberystwyth. The first major show case for Llanarth Flying Comet to show his paces under the new owners was at the Royal Welsh Show held in July. Following a tense afternoon of judging, Gareth Evans, the judge, whittled the final line-up to two cobs, eventually awarding the Prince of Wales Cup to Llanarth Flying Comet. Pauline, sitting in the stands smiled with pleasure, an affirmation of the quality she had entrusted to the new owners. Such was his success in show rings up and down the land, he went on to the Horse of the Year Show, winning the championship at Wembley in 1979 and 1980 – a handsome accolade for the Llanarth stud breeding policy.

Len as partner and a knowledgeable expert cob breeder and handler ensured that the show cobs went to all the major events not only during the hot summer of 1976 but in the following years, although the college authorities looked askance at the financial implications of entrance fees, special feeding, travelling and stabling costs. Pauline and Len knew the importance of keeping the best cobs in the public eye, especially to attract breeders and potential buyers. And Flying Comet, like Braint, knew how to draw applause from show ring grandstands – head held high, ears pricked, eyes flashing, coat gleaming, and the long, supple, flowing action – the epitome of strength and grace. Publicity and promotion was a vital ingredient at showcase events, and when spectators were allowed to vote in a new competition – the popular applause competition, Flying Comet, living up to his name, won 'hands down'. Pauline basked in the adulation as she witnessed the perfect partnership of Len Bigley and Flying Comet stepping out and showing their paces.

The sale held in the autumn of 1976 attracted many overseas buyers and was considered the most encouraging despite being the thirteenth in the list of Llanarth sales. The catalogue noted the new owners as the UCW Aberystwyth – the University gained valuable publicity from all the honours heaped on the stud during their first show season. But, as the year drew to a close, tensions surfaced. The farm manager needed to change the emphasis, to invest and improve the land. The farm had been run for the benefit of the horses, at least forty or fifty of them and that number of cobs could easily ruin pastures, by cutting it up in wet weather. Pauline regarded the stud as a priority, the horses should have the best of everything but Jim Lees had other pressures – how to make the farm more efficient, to reduce costs and to make the farm pay.

Her reliance and trust in Len and his respect and loyalty to her made his position difficult. His priority as it had always been was to ensure the stud and its breeding policy was maintained but as he remembered:

> The College invested heavily in the drainage operation. It certainly needed to be done, much of the land was bog and needed to be improved and made productive. But they were spending vast sums of money – unlike any other farmer – and the return on that investment would take years to recoup.

Old practises were questioned especially costly ones, but Jim Lees did manage to purchase a new horse box for the stud but the small matter of obtaining a fire extinguisher to comply with health and safety regulations suspended belief and became a matter of wonderment. Len recalls, 'A man came to view the farm and buildings. The following week another man arrived to decide where the extinguisher should be placed. The third time a man and a mate came down to fix it, and they said another would come to see if it was in the right place.' An apocryphal story like that gathers its own momentum but there was merit and certainty in a recollection of the farm petrol pump:

> We had a petrol pump in the yard. Pauline always had one there. The farm paid for it. We filled up when we needed to. When the University came they couldn't believe there was this pump and no one keeping a record. A council road ran between the house and the farm buildings and in a way you didn't know who was filling up so I fully understood the need for tightening up on all that and many other things.

Events during those first two years were an ominous prologue to the changes that corroded relationships during the next three years. It began when the University College restructured the Agriculture Department's managerial structure, taking in the different requirements for academic teaching and research and the management of eight farms. Professor David Morris was to direct the management of the college farms and the needs of students in their practical courses and Professor Desmond Hayes took responsibility for the academic side and research projects. It was an amicable sharing of leadership duties and responsibilities.

But it heralded a clash of ethos and an upheaval for Blaenwern and the Llanarth stud. In order to make the farm financially viable, Professor Morris had decided that the farm, which had already received a large investment from the

College to improve the land, should become a commercial farm. He wished to change the emphasis of the farm's husbandry and management of all stock and to review the number and types of animals kept. Only productive Welsh Black cattle would remain and they would be wintered and fed indoors; the flock of Welsh sheep would be crossed with another breed to produce lambs for spring sales and the number of cobs in the stud would be considerably reduced so that grazing land for cattle and sheep could be managed effectively. It was a policy which would disregard the intention to create a centre for breeding pure native Welsh breeds.

Dr Morris met Pauline and Enid for the first time to outline his new plans. The meeting was courteous but the message was unexpected. They were unprepared for such sweeping changes and when he left the house they felt frustrated and fearful. Len Bigley remembers the outcome vividly:

> That first meeting when he came out of the house, the disrespect he showed, the comments he made about them when he told them what he was going to do. He may have been right in his intention of the things he was going to do, but it was his manner. He was probably tactful in the house, but to us he declared, 'I'll get rid of it all'.

They were two elderly ladies who had gifted their home and 'living' to a prestigious institution and now they began to feel that their intentions for the future were being cast aside. Indeed, it seemed that forty years of work would be dismantled and aspects of the agreement threatened. During their conversation there had been no hint of setting up a national stud for Wales or that there would be fewer cobs of certain bloodlines to remain in the stud.

The year had begun with hope and it ended on an apprehensive note, although there was little sign of their true feelings in their Christmas card greetings. Every year since she left Blaenwern, Barbara and Pauline had exchanged greetings on cards which doubled up as a promotional advertisement for the stallions at the stud. Few of these cards remain, but the 1976 issue sent out to breeders shows a photograph of Llanarth Meredith ap Braint in full high stepping stride, his rippling muscular strength and speed caught in a split second by the camera. Details of his pedigree and costs set out: fee £40.00, groom £1.00, as it was for the other stallions listed: Llanarth Meteor and Llanarth Flying Comet but the fee for Llanarth Braint was ten pounds higher at £50.00. An earlier card written when she had recovered from her broken hip apologises in her inimitable style for the delay:

> I did not send you a goose this year as I thought you might not be able to deal with it! I am late sending out many of my cards as I have not long come out of the hospital at Oswestry. The operation has been a great success and I am making good progress. In some weeks I expect to be walking again without aids – and maybe to ride again in time. It is wonderful to be free of pain and stiffness in my hip.

The card for 1976 makes clear that the ownership of the stud has passed to the University College of Wales, Aberystwyth and Pauline writes to Barbara of her relief at the agreement:

> The UCW are now in possession of the farms and stock, as you see, no more financial anxiety, thank God! We stay on just as before and Enid keeps the house here and Penrhiw till we depart this life… All is on condition that it remains the permanent home for the Welsh cob and cattle. Have just returned from a marvellous four weeks in USA and Canada.
>
> Love and best wishes for Christmas and New Year,
>
> Paul

During the next two years Pauline and Enid were mindful of the proposal to disperse another world-famous Welsh stud, the Coed Coch stud at Abergele, north Wales, which had been established in the 1930s by Margaret Brodrick, the great ambassador for Welsh mountain ponies. The estate passed to Lt Col E Williams Wynne, but almost twenty years following her death in 1962, a scheme was proposed to purchase 35 ponies of certain bloodlines and house them at the University College of North Wales, Bangor. The financial implications of such a scheme were prohibitive, the proposal was refused and as a result 244 ponies were sold at the dispersal sale on the 7th of September 1978. For forty years Daisy (as Margaret Broderick was affectionately called) together with Shem Jones her loyal groom had developed a unique stud. It proved to be the finest collection of native ponies in the land and as someone commented, the Coed Coch champions were, 'as thick as daisies'. It seemed appropriate that the last pony to be sold was called Coed Coch Pen-y-Daith (Journey's End). It was a salutory reminder for Pauline and Enid that there was little permanence in stock breeding.

But there was one celebratory occasion which could not be ignored. Llanarth Braint had reached the grand old age of thirty in May 1978 and the College organised a worthy celebration in his honour. He was the only member of the stud to remain under Pauline's ownership and to mark the occasion, two marquees were erected near the house with one of them displaying all his rosettes, cards, cups and plates in addition to photographs of all his descendents. The centrepiece in the other was two magnificent birthday cakes. Braint's cake was a confection of carrot and apple with a model of a horse rearing majestically from the covering of icing. The other cake had a magnificent model of a palomino, also on a bed of icing to note Braint's chance inheritance. A crowd came to mark the occasion and the high spot of the afternoon was Braint's entrance along with his constant companion, Alphonso the donkey, who kept well in the background. Braint knew instinctively how to react to the welcome and the applause – his eyes bright, ears cocked and a spring in his step and, as always, he kept that 'sense of occasion' – he didn't eat the cake while on parade.

The 1978 Christmas card included a photograph of Llanarth Flying Comet, with the familiar list of stallions: Flying Comet, Meteor, Meredith ap Braint and Braint himself, and that year, Pauline wrote to Barbara of their travels to shows in Germany and Austria ending her note thus, 'The UCW have booked a special niche in the Equestrian Fair for Llanarth – so I just may be there too. Both Meredith and Meteor are going to perform, I believe.' There is a hint of uncertainty in her role as the consultant to the stud, but there is no mention of the ever growing doubts and anxiety in her card a year later, 1979, which has a photograph of Meredith ap Braint's noble head as a frontispiece. Her carefree attitude to driving and to other drivers seem to hide hidden concerns and to divert pointed questions about the stud and the farm, 'Enid and I were in London last week for a few days. Luckily, still find driving the car quite easy and traffic does not bother me.' She goes on:

> We went to the Royal Philharmonic concert and next evening to a most interesting film, *Light in the West* (second in planned series of 13 films), originally based on my father's book, *The Coming of the Saints*, written and published in 1907, it has been republished and has a new lease of life! Unless some miracle happens I agree with you, our civilisation seems drowned through sheer greed and selfishness.
>
> With love and best wishes from Paul.

But despite the recession that Pauline obliquely refers to, the Llanarth sale attracted over 3,000 people and was a resounding success. That year the cob supply did not match demand and a young researcher, Patricia Aidley, who submitted her research in the form of a dissertation for an MSc degree concluded, 'Cob sales are a reflection of the physical, economic, cultural and historical isolation of the Cardiganshire men involved.'

I'm not sure whether her conclusion of isolationism is correct because a demographic analysis of those who attended the sale showed a high proportion, 44 per cent were English and 50 per cent were Welsh. Preparation for the 1979 sale was filmed for a BBC 2 series, *A Diary of Britain*. Called 'Cob Country' it featured many studs within the close knit community, and of Pauline and Enid entertaining fellow breeders around their dining table and, as the wine flowed, the conversation on the merit of the cob became animated and intense with Pauline at her most forthright.

But 1979 will be remembered as the year that Braint collapsed and died in January at the grand old age of thirty-one. He was buried under the old chestnut tree in the paddock and a memorial trophy was presented to the Royal Welsh Agricultural Society in his honour by Pauline and Enid.

CHAPTER ELEVEN

T HE NEXT TWO YEARS were cataclysmic for Enid and Pauline. The agreement with the College was disintegrating and no one seemed to be listening to their point of view. Professor Morris, following his agenda of making the farm financially viable, was given full backing by the College authorities despite mounting debts and interest rates 'going through the roof'. Sir Goronwy Daniel and Sir Ben Bowen Thomas had retired as Vice-Chancellor and President respectively and their successors, Dr Gareth Owen and Lord Cledwyn Hughes faced the urgent task of making economies. Farm accounts were studied in great detail, as was expenditure in all other university departments, but the substantial investment programme at Blaenwern to improve and modernise systems and practises, meant considerable losses month by month.

Professor Morris pressed ahead, and perhaps one incident above all others seems to sum up the revolutionary changes already underway. He ordered that all Welsh Black cattle be dehorned to minimise the risk of animals causing injury to each other as they proposed a change of practise. The breed is characterised and noted for their long horns but to improve efficiency and productivity, changes to husbandry practises, such as keeping and feeding the cattle indoors during winter months, minimised the dangers. Dehorning older animals was often seen as barbaric but today breeders are able to breed naturally polled cattle from those who carry the latent inherent gene. The Welsh Blacks had always been Pauline's pride and joy. Anne Wheatcroft remembers the incident, 'She adored them. Pauline came back to the farm one day, heard a commotion and a lot of noise and went to see what it was. They were dehorning her cattle. They had never mentioned a word to her.'

Pauline was heartbroken. She found the scene horrifying and distressing but there was nothing she could do. She and Enid talked long into the night voicing their concerns for the future and realising that their vision for Blaenwern becoming a centre for nurturing native Welsh breeds was crumbling. Slowly, they formulated a plan of action. First they needed to find a solicitor who would act solely on their behalf. They needed an independent legal mind to review the clauses in the document – were there any loopholes or points of weakness? Trust and respect were ebbing away. Enid felt that she could no longer believe that the College would maintain the estate in its original form and, as her disappointment deepened and sharpened, she made it clear that she had no intention of discussing anything further with Professor Morris.

Anne Fowler, a feisty legal mind, was immediately engaged as their solicitor. At that time she was working in the Midlands but she had first met Pauline twenty years earlier through her interest in horses. The first stallion she bought was called Llanarth Cerdin. Pauline recognised him immediately at one of the shows and commented with a smile of satisfied recognition on her face, 'What a beautiful stallion you have.' He was a winner at several important shows and he was described in the first Llanarth sale catalogue, 'Cream stallion. Pony of cob type. Remarkable for conformation and action. Excellent stock getter. Quiet to handle and has been ridden.' As if to prove his qualities, the bidding rose to £475, but Pauline withdrew him from that particular sale.

Anne Fowler had made her home in readiness for her retirement in a whitewashed cottage in the village of Llanbedr, three miles to the south of Harlech on the shores of Cardigan Bay. They were to meet up at other events and shows and by the end of the 1970s, when matters at Blaenwern were going 'horribly wrong', Pauline sought her professional advice and Anne Fowler at their first meeting found the ladies nervous and distressed, disappointed, worried and unhappy.

'They were desperate old ladies. As we talked, I could see she was living in cloud cuckoo land at the time because she could not understand why her wishes were not carried out.' They had considered suing the College for the return of the farm and Len had begun proceedings against the College over his own job. The local press reported:

> Threats of legal action by the former owners of the top cob stud against the University of Wales. There has been controversy between the former owners and the college management culminating in a clash of personalities and the threat of legal action. Towards the end of 1979 Miss Taylor raised objections to the way the farm was run resulting in a meeting with the college authorities.

The College had realised that to ease communication and understanding between both sides, they had to establish an advisory committee to assist with some of the problems, especially those involving the stud. Anne Fowler and Pauline's nephew, Tony Bradshaw became co-trustees, along with Len Bigley and Wynne Davies representing the Welsh Pony and Cob Society. Pauline announced, 'The situation is now much more hopeful and it looks as if things are going to be resolved.' The Principal, Dr Gareth Owen made this comment, 'There have been some problems and I am hoping that these are behind us now.'

But there remained legal matters to resolve and those included safeguarding Len's future because Pauline was determined to look after his interests and also to keep the stud intact. Enid, who kept her feet firmly on the ground, was anxious that Len should inherit the house, but their wills stated that they were bequeathing everything, including the house to the College. Enid had, at one time, considered leaving one of her original cottages, Penlon, to Len but the legal implications of the agreement at that time was too complex. Len had been an integral part of the stud's development but the College was taking away his rights, neglecting his knowledge and without once describing him, the stud manager, as an asset. Anne Fowler was asked to examine the possibilities and to draw up new wills, 'Enid and Pauline wanted to protect Len. They regarded him as a son and he was now courting Ann, 'a low key affair but serious', and they were anxious that the young couple should inherit some land with the house. The college refused.'

Len and Ann had met at shows and events and, although she was brought up in the south of England, she spent all her holidays with her grandparents on a farm in Dihewid, Cardiganshire. It was there that she developed a keen interest in riding Welsh cobs. The Hewid stud was well-known and respected because one mare, Chancerie Polly, which they bought in 1963, became a champion in the show ring and perhaps, more important to breeders, her bloodline was to influence the development of the breed worldwide. Len and Ann married in 1975 and went to live in the cottage Penrhiw on the estate for a year and then bought a cottage, Esgaironw near Talgarreg. These were uncertain times as they began their married life with deepening worries for the future. Pauline and Enid signed their changed wills on the 1st of April 1980 leaving Blaenwern house to Len and appointing Anthony Bradshaw and Anne Fowler executors and trustees.

In the spring of 1980 the President of the University College informed the council:

> ... that there was at present a disagreement about the way in which the terms of the Charitable Trust, set up after Blaenwern was given to the college, were being carried out. The officers were in touch with the donors and hoped that they could resolve the difficulty. He would if necessary inform Council of any further developments.

This was the first official minute about the conflict and wrangling at Blaenwern. Pressures were mounting. The College was looking for economies and productive management; the ladies of Blaenwern were seeking to protect their vision for the estate and the stud. It was neatly summed up for me by Professor Desmond Hayes:

> David Morris had to explain the changes and improvements necessary for production on a commercial farm and how they were going to deal with it, and remember, the ladies had always run their own business, they had never taken orders... two cultures, two philosophies... one being in control, the other, this is what needs to be done to establish a commercial farm and new practises alien to Blaenwern were to be introduced.

Three months following the Council meeting, Enid died on the 19th of June 1980, within two weeks of her ninetieth birthday. She had been picking gooseberries in the garden, and had then set about making a few pots of jam which she showed with pride to Pauline and Anne Fowler. 'I'll go and make a cup of tea for the tasting,' she said and disappeared to the kitchen. 'It took a longish time, and when we went to the kitchen, she had collapsed and was lying on the floor. We called an ambulance but Enid had died suddenly and instantly.' Anne Fowler recalled, 'She picked gooseberries, she made something that would last, and then left.'

Pauline was distraught. She felt responsible, but they had both realised that, as two elderly ladies, time was running out. Their views counted for very little. All they had achieved for the Welsh native breeds in the past was being eroded and lost. It was painful and distressing. She summed up her thoughts with these words:

> It is a sad and bitter thought for me to know that Miss Lewis died without knowing the outcome of this controversy, sadder still to reflect that her last months were made so unhappy by something she had thought would bring pleasure, renown and profit to Wales.

Enid's funeral service was held at Beulah Chapel in the nearby village of Talgarreg and was followed by a private cremation. Within months, a concert in her memory was held in the Old Hall arranged by the Department of Music at University College, Aberystwyth and donations were given to the Musicians Benevolent Fund. It was a moving occasion, the programme included two compositions by Hubert Davies, 'Antiphony' (1941) dedicated to Walford Davies, and 'Allegro Scherzando', dedicated to The Dorian Trio (Kathleen Washbourne, Pauline Taylor and Enid Lewis). There was a solo from Gwawr Owen (piano) 'Prélude, Choral et Fugue' and baritone Delme Bryn Jones sang the aria 'O Star of Eve' from the opera *Tannhauser* and the Welsh ballad 'Berwyn' by D Vaughan Thomas. Delme was at the peak of his career and since retiring to Blaenwern, Enid had devoted a great deal of time assisting him (he was a former student at the Guildhall now living in Llanarth) with his work and offering support as he tried to recover from his addiction to alcohol. He won acclaim from the world's major opera houses although occasionally there was a feeling that he was too cheerful-looking to sing the villainous parts often given to baritones. Gwawr had also lost a wise mentor. They corresponded regularly when she began her first term at the Yehudi Menuhin School:

My dear Gwawr,

It was so nice to get your letter telling me all about your timetable etc. – and by now I hope your cold has completely gone. You may have heard from Mrs Chapple that we spent a little time in London and had coffee with her at the Festival Hall last week. She gave us first hand news of you and said that you were settling down and were happy. I was interested to see what new piano work you have been given – you'll have to work! It's good that you are to play to Louis Kentner and have a lesson from him. When we were in London, we heard the Leningrad Orchestra and on Tuesday last we heard the Royal Liverpool Philharmonic Orchestra at Aberystwyth. Tonight we are going up again to hear Christine Onlij play – I have never heard her but I understand that she is a very fine pianist. What a feast of music in a short time! Saturday week is our annual sale and the BBC is doing a feature on this place and we understand that it will be shown on BBC 2 next year. They are spending days here, so let's hope it will be good. I miss you very much, but Mr Nichols has you now instead and you will do very well with him. With love from Miss Taylor and me.

Yours sincerely, Enid Lewis

The letter was written in October 1978, the writing firm, and the sentiment warm and encouraging. In her will she bequeathed most of her music to Gwawr and the personal letters and cards are treasured mementoes of a very special teacher. One of the last was a 1979 Christmas card – a photograph of Len Bigley and Llanarth Flying Comet receiving a silver plate from Princess Michael of Kent as winner of the Supreme Champion at the Horse of the Year Show.

But they were dark days during the first summer of the new decade. Enid and Pauline had formed a long partnership in music and farming and, during the last fifteen years of a sixty years of friendship, Enid had invested her life's savings into recreating the Blaenwern estate which had been her grandfather's legacy. She had backed Pauline to maintain the Llanarth stud and to establish a centre for native Welsh breeds. But that mission was in danger of being obliterated.

* * * *

Len and Ann, with their new born baby, Simon, moved into Blaenwern farmhouse to care for Pauline, her spirit and will to live weakening as the months passed. Changes at Blaenwern moved apace. Professor Morris seemed to sweep away practises without consultation and the situation for Pauline had become very different from anything that she had imagined. She spent her last weeks in hospital at Aberystwyth. Her friends ensured that she was never left alone to dwell on her deep unhappiness. Her thoughts focused on her horses and what the future held for the stallions and what was happening to her Welsh Black cattle. Anne Fowler remembers the agony of her torment, 'The farm should form part of the heritage of Wales. She told me that time and time again. They wanted everyone to benefit, to share in what they had done.'

Following a further year of arguments, when legal proceedings over the new wills were agreed, Pauline died on the 17th of June 1981, almost a year to the day of Enid's death. Tributes from leaders and friends of the cob and farming world were fulsome in their praise. Tom Owen, the College Registrar, sent a gracious letter to Tony Bradshaw, Pauline's nephew:

> As you are the closest member of her family known to me, I am writing to you to record my sorrow at the death of Pauline Taylor. As you know, we had a great deal to do with each other over a long period, and my life was enriched by this. Such a multifaceted and robust personality could be endearing and fascinating. Without Enid and Pauline, Blaenwern will hardly seem the same.

Betty Davies, a long-standing friend, who farmed TyGlyn, Ciliau Aeron, said at the time in answer to the question, would the ladies have lived longer if there had not been such acrimony? 'It's something that we, her friends find very hard to forgive, the fact that the vision was broken in front of them. There was not a real understanding of the meaning of the gift.'

Pauline's funeral drew large crowds, her ashes were later scattered in the peace and tranquillity of Blaenwern gardens. Betty Davies wrote a more formal tribute in the *Cambrian News*, the weekly newspaper for the area, ending with these words:

> In 1961 when Barbara Saunders Davies moved to London and the partnership ended, Pauline, showing typical courage and determination, decided to carry on. She never minimised the strains and difficulties she underwent and she always gave full credit to

Barbara's considerable genetic knowledge which gave the stud as good a start. But it was from this time, under her direction that the stud began its steady rise to world renown and a success... Pauline Taylor had complete integrity of outlook and was an indomitable fighter for the things she believed and always ready to share her knowledge and experience. She cherished old customs and values but was always ready to give new ways a trial.

William Lloyd remembered, 'When I started breeding on my own she gave me every encouragement and support. In fact, she was very supportive of young people. She always looked to the future, rarely to the past, often asking the question how are things going to develop.'

A report by the Committee established by the College Council on the 9th of July 1980 with plenary powers to consider matters relating to Blaenwern, was submitted to the Council at its meeting in July 1982. It was received and adopted:

The Committee is pleased to report that a compromise has been achieved to the dispute with the personal representatives of the late Miss Enid Lewis and the late Miss Pauline Taylor and that the terms extend to Mr and Mrs Len Bigley and Mr Northam (the stud groom).

It will be remembered that following an approach made to Sir Goronwy Daniel by Miss Lewis, arrangements were made under which she gave Blaenwern farm (but not the house) to the College on defined charitable trusts. Miss Lewis and Miss Taylor then sold the Llanarth stud to the College but left part of the price on mortgage. As part and parcel of the deal Miss Taylor, Mr Bigley and Mr Northam were employed by the College as consultant, stud manager and stud groom respectively. The College also undertook certain obligations in relation to Blaenwern house which it was understood would, in defined circumstances, come to College with a cancellation of the mortgage on the death of the survivor of the two ladies.

Unhappy differences arose in 1980 following which Miss Lewis and Miss Taylor asserted title to the farm and stud and changed their wills. For a while during that summer Mr Bigley and Mr Northam, acting on advice, declined to take orders. These events were overtaken first by the death of Miss Lewis in June 1980 and then by the death of Miss Taylor in June 1981.

After the death of Miss Taylor, Mr and Mrs Bigley moved into Blaenwern house and it is understood that Mr Bigley had been promised tenancy.

Following the death of Miss Taylor, the College commenced proceedings to enforce the promised bequest of Blaenwern house and cancellation of the mortgage. These proceedings were contested by the personal representatives who themselves commenced proceedings to enforce the mortgage.

Early this year, the College received an approach for a wide-ranging compromise. Exhaustive negotiations then took place as a result of which it was eventually arranged that the mortgage would be cancelled. Mr Bigley and Mr Northam would leave without payment of compensation, the personal representatives of Mr and Mrs Bigley and Mr Northam would refrain from challenging the administration of the charitable trusts affecting the farm, the personal representatives would convey three areas of land surrounding Blaenwern house to the College, the College would withdraw its claim to Blaenwern house and part of the immediately facing land, the College would pay £2,500 to Mr and Mrs Bigley and both sets of proceedings would be withdrawn on the basis that each side pays their own costs.

We considered most carefully the advice received from Counsel, a report from an independent firm of valuers and the complex documentation prepared to give effect to the foregoing terms and we came to the unanimous decision that the compromise should be approved.

Accordingly, acting under our plenary powers we authorised the Principal and the Registrar to take the necessary steps to implement the arrangement.

CHAPTER TWELVE

T HE COMMUNITY OF LLANARTH and the wider pony and cob fraternity had become exercised at the unravelling of the Blaenwern legacy. Rumour and gossip had found fertile ground, with the College refusing to reveal their plans for the future. The prolonged legal wrangling as a result of change in the bequest contained in the wills and its effect on the agreement between Pauline and Enid and the College was delayed, but eventually agreed. Len Bigley and his family inherited the farmhouse, much of the furniture and personal possessions. His position had become untenable; he was without authority and was ignored by managers and senior College authorities who had sidelined the importance of the stud and the practises of breeding, selection, training and promotion. The notion of creating a national stud as part of the Welsh heritage was suppressed and many believed that the obligation placed on the College to create a centre for research and training students in equine studies was not discussed.

The future for Len Bigley and his wife at Blaenwern was bleak; their only course was determined by their own aspirations for the future and fulfilment as a family. They began searching for a farm to establish their own pony and cob stud. That search took them three miles over the border to England, eventually settling in the parish of Michaelchurch Escley, near Hay-on-Wye, where they hoped for a measure of peace and tranquillity on a farm called The Quakers Farm. This was the place to start again, to set aside the bitterness and the disappointments of the past, and to accept the challenges of putting down new roots. As its name suggests, the house was a Quaker meeting place, built in the seventeenth century between the Black Mountains and the Golden Valley. The 140 acre farm of mainly rich, red sandstone loam coupled with a temperate climate was ideal for good grassland and making quality hay and silage. For eighteen years Len had been involved with the marginal lands of Blaenwern and the development of one of the renowned studs in Wales but that 'hoped for' continuity had disintegrated. It was now 1982.

Six years after accepting Blaenwern as a gift, the College authorities began discussing the very real possibility of selling the farm, the animals, machinery and equipment. This was when the loosely worded clause in the initial agreement, 'that the farm could be used for the general purposes of the college', became operational as the last loophole within the legal document. 'I feel bitterness and resentment', was Len's reaction at the time, 'They have let Pauline down very badly and on reflection, I don't think they intended to carry out her wishes, I really don't.'

In September 1982 the Farms' Advisory Committee reporting to the College Finance and General Purposes Committee recorded, 'The problems at Blaenwern are well-known but the way was clearing now for decisive action. The Farms' Manager indicated that the estimated figures for Blaenwern were generally achievable, with the possible exception of stud labour costs.'

That report went before the Council at its meeting in July when the recommended 'decisive action' was agreed: to sell the farm, the animals and equipment but to keep 16 cobs of particular bloodlines as a core stud to be housed at another college farm. The College had incurred debts amounting to £300,000 in six years. They had invested heavily, in roads, drainage and new stock and the College Registrar, Tom Owen, explained that their main financial effort went into capital investment at a time when interest rates were going through the roof. The farm could not sustain the huge burden of such a debt, 'No way could we go on.'

New year 1983 began with rumours spreading like wild fire but the College refused to make an announcement. Cynog Dafis, prospective Plaid Cymru candidate for Cardiganshire wrote to the Principal in April asking for clarification as many people were already enraged that such a famous stud and farm should be dispersed and sold. A few believed the farm had already been sold, others offered suggestions on how the stud could be saved since the income from that part of the business in 1981–2 amounted to £26,000. In his letter, Cynog Dafis alleged, 'It is all but incomprehensible how the College had managed to transform a valuable asset – a 380 acre farm given for nothing and turned it into a financial burden.' He called for a detailed inquiry under the auspices of the College Court into the running of the farm. Wynne Davies had already put forward a proposal that the College could maintain the stud, together with sufficient land, placing a young farmer in charge of the enterprise. The following week, the Farmers Union of Wales called to see the balance sheets. It had been suggested that that the farm's annual bill of interest on borrowing was £27,000.

Eventually, the College issued a press release confirming the proposed sale of Blaenwern farm and the stud. The farming unions were appalled. One Cardiganshire member of the National Farmers Union at an executive meeting in April described the press release as, 'A load of bunkum. It really tells us nothing. The truth is not coming out, the press release gives us no figures.' Another member was aggrieved to learn that the stud would be dispersed, 'Going back to the time when it was handed over to the College, was it not to safeguard Welsh breeds? We should now think of the safety of stock lines, we need to carry on the good name of the Llanarth stud.' Perhaps the most telling remark was from a member who questioned the competence of the College, 'What I think is important is that the farm is in a marginal area. The money spent and the work that has been done there is vitally important to all of us.'

Petitions and letters to the local press added to the steam erupting from the Cardiganshire cob cauldron. Betty Davies took issue with a statement that implied that the farm, when it was gifted to the college was financially unviable and debt-encumbered. She wrote:

> Of course Pauline had an overdraft. It would be interesting to speculate how many farmers have got through their careers without one, and I remember clearly her satisfaction and pride the day she told me that she and Enid had just been with their accountant and the farm account was at last in the black... It has been stated that the College paid for the stock. I would like to explain that: (i) No sum of money was handed to either lady. (ii) The valuation was a very low one. (iii) It and other arrangements were mainly an accountant's perfectly permissible way to ease any tax situation and also to make provision for annual payments to both ladies. Pauline had been appointed consultant to the stud. It must be appreciated that both were elderly and were handing over both livelihood and capital and naturally were not expected to exist on nothing for the rest of their lives.

Another letter in the *Western Mail*, written by G J M Thomas, an old student of the College, firmly believed that the stud was included as a gift in the hope that students of agriculture would benefit. 'There are more agriculture students than in any other science subject,' and he goes on, 'Finance can be a limiting factor in farming, as I well know, but one tries not to dispose of one's most valuable assets and above all land. Surely staff of the Agriculture Department can meet the challenge of financial problems and demonstrate to students how such situations can be faced.'

A letter dated 6th of April was written by Enid's first cousin, Gwyneth Jennet Margaret Thomas, living in Nelson, Pontypridd. She and Enid were like sisters, brought up close together in Pontypridd and throughout their lives they were very much in touch. It was the unanimous view of the family that the Blaenwern estate and the stud should have been left to Len Bigley but Enid and Pauline were persuaded by the College Principal during their early discussions that the farm would be safer in College hands.

The Farming Unions continued to voice their opinions until the week beginning May 1st when a headline in the *Cambrian News* proclaimed:

FARMERS URGED TO HOLD FIRE OVER LLANARTH STUD ROW

Farmers were urged not to be too critical of UCW over the stud until they knew the full story. A Cardiganshire county executive of the National Farmers Union had heard from a member who was also a member of the Council that he was unable to say much because of 'the confidential nature of the issue'. Another member added, 'Now I feel what we should be contributing is advice. But everything is submerged in a blanket of secrecy.' Another said he didn't believe any amount of advice would save the College debt. Cynog Dafis, who had not received a reply to his letter, called for the College authorities to make a statement to scotch the rumours that Blaenwern had been sold.

Four days later the Principal made a public statement categorically denying the rumour that the Llanarth stud had been sold. 'The college is well aware of the points raised by Mr Dafis. They are actively under discussion.' The Principal also denied the suggestions that an injunction had been served on the College preventing them from selling the stud.

The independent valuation of the Blaenwern estate when the agreement was signed in 1976 reflected the state of the country's economy. It was extremely low – £65,000. Five years later the College asking price for the estate was in the order of £425,000. The journalist Bruce Kent conducting an interview for a television programme into the Blaenwern saga, asked the College Registrar, Tom Owen, 'Where did the money for the investment at the farm come from?' And Tom Owen replied, 'The money came partly from the sale of another farm which we owned and partly by way of a loan. We now had a huge burden of debt, in the region of £300,000 which the farm could not sustain.'

In the week of the Royal Welsh Agricultural Show, the third week of July 1983, the College issued a statement:

> UCW has informed the Farmers Union of Wales that it intends to proceed with the sale of Blaenwern farm. The farm will be advertised for sale. Regarding the Union's concern about the future of the stud, the College is already in discussion on the possibility of retaining a few of the Welsh cob bloodlines and they will be housed following the sale on another farm owned by the college.

The majority were shocked at the scale of the £300,000 loss and that such a large sum was accrued in six years – a short time to accumulate debts of that magnitude. The College Council accepted the recommendation and decided that the sale should proceed, but the farm was a separate issue from the stud. The College sought the advice of the Welsh Pony and Cob Society, asking a prominent breeder, 'You are the expert, which blood lines would you recommend we keep?'

Fourteen horses, two stallions, seven mares and four yearling fillies were selected from the stud of thirty horses, a reduction from the number the College bought when they took over. Following the Blaenwern sale, the stud of fourteen would be housed at Frongoch, the College farm two miles to the north of Aberystwyth where more money would be spent adapting buildings and building new ones to accommodate them.

The nineteenth Llanarth sale in October 1982 was a sad ending to the last of the Llanarth collective sales. It was not a typical St Luke's little summer's day – the atrocious weather conditions seemed to reflect the wrath of all the saints, indicating their disapprobation and admonishment at events about to take place at Blaenwern. It affected prices. Only 81 out of a total of 154 horses found new owners, averaging a price of £272, in total £22,097 – the lowest figure for ten years. Interest and confidence had plumbed to the depths reflecting the mood of the time and the attitude of breeders.

The following year arrangements were made for the sale of the estate which would be organised by tender once the valuation had been agreed. The stud, other stock and machinery would be sold by auction. Reaction to the news rumbled on and the Farmers Union of Wales issued a statement at the Royal Welsh Show in July, 'Confirmation of the College's intention to sell would dismay cob breeders. The farm was given to the College in order to protect the future of the stud and other pure breeds of Welsh livestock.' But cobs from the Llanarth stud were notable absentees at the 1983 show, a stud which had, over three decades used the event so effectively to promote the versatility of the Welsh cob. It was a bitter irony that it was also the year when the new permanent building for the Welsh Pony and Cob Society at the show ground was officially opened by the Queen, followed by a parade of sixteen stallions, four from each section.

In September an advertisement appeared in the *Cambrian News*:

HIGHLY IMPORTANT SALE

Llanarth Pedigree Welsh Black herd,
Welsh Cob stud Sheep and Implements

at BLAENWERN, LLANARTH, DYFED

Saturday 24th September, 1983

Sale to commence at 10.30 am

148 registered and non-registered Welsh Black cattle
12 registered and 4 non-registered Welsh Black calves
29 registered Welsh cobs, Section D
10 mares with foals, 7 yearlings and 2 foals
Finest examples of breed and very interesting pedigrees
with the famous Llanarth bloodlines strongly represented
1200 ewes, lambs and rams
Comprehensive range of machinery

A week earlier the Blaenwern estate of 380 acres was sold. It was bought for the asking price of £425,000 by Mr and Mrs J O Evans of Pansod Farm, Synod Inn who already owned large tracts of land in the area.

CHAPTER THIRTEEN

A CROWD OF THOUSANDS gathered to witness the final act in the turbulent events at the Llanarth Welsh cob stud. Fourteen animals retained by the College were already at Frongoch Farm but on a dull Saturday in September, 25 cobs faced the auctioneer's hammer. Many breeders and farmers were angry and bitter at the decision to sell – a mere six years since the College had received the farm as a gift. It was a feeling that was summed up by Anne Wheatcroft who strode majestically without warning into the sale ring:

> Ladies and Gentlemen, [She had no need of a microphone] Miss Taylor had a vision, a dream that one day her cobs would belong to the Welsh nation. Now sadly, very sadly, it is coming to an end, as all things must. And now I'm afraid her vision is turning into a nightmare. I trust we will salvage a little of what she's done. Thank you.

The auctioneer was apoplectic because he had cautioned her that no-one was to speak before the auction, it was too controversial, but she went on the attack, 'I was so angry. The last sale at Blaenwern. It was a horrible drizzly day. There was a long pause at the beginning because they couldn't get the first cob into the ring – there were people handling them who had no experience – and before I knew where I was I was in the ring.'

The first cob to be sold took centre stage and within an hour, the Llanarth cobs – 25 mares, foals, fillies and geldings had been dispersed to new owners and for many it put into perspective the complex and demanding process of creating a premier stud. It had taken four decades to seek perfection, and just an hour to sell and disperse. Bidding in the gloom had been brisk and the total amount for the equine section was £15,084, at an average of £603 per sale. Llanarth Sue

Ellen, a four-year-old mare topped the sale at 1,700 guineas. When the cattle, sheep and various implements had been sold, the sale had raised well over £100,000 and when that sum was added to the £425,000 for the estate, there could be no complaints. But there was sorrow, anger and a feeling of desolation around the sale ring.

The College announced that the Llanarth stud would live on at another College farm but two months later, when the Principal reported to the Council, he was much more cautious with his use of words. The financial situation remained unclear and proposals for the stud's future as recorded in the minutes were equally unclear. He went on to say, 'In the meantime the nucleus of the stud has moved to another College farm as an interim measure.' At that meeting, a council member referred to the unsatisfactory manner in which the issue has been dealt with by the press and the broadcasting media. Also, he expressed his concern at the prospect of the College retaining part of the stud, if their retention was at the expense of academic posts. The Principal reaffirmed that it had been agreed as an interim measure, and that future policy regarding the stud would be decided when the financial picture, following the sale of the farm and stock was clear.

Anger and exasperation erupted in the cob world. Only a few breeders sent their mares to the stud at its new location and the financial problems intensified. Eighteen months later the College Council decided to terminate its responsibility for the Llanarth stud and to sell the remaining animals. Gareth Owen, the Principal, taking part in a television documentary in May 1985 commented, 'We took the decision to sell because of the cost of running the stud. We couldn't afford those costs which were in the order of twelve to fifteen thousand pounds a year to maintain.'

The financial situation in higher education had become critical and a government Green Paper revealed proposals for major financial cuts at every university in the UK, and to save money at University College, Aberystwyth, the remnants of the Llanarth stud was to be sold. But it was not the only activity under threat. The Principal identified other academic departments and core subjects which could be cut – a severe reduction of the College's 32 departments. The three most threatened were Chemistry, Philosophy and famously, the Music Department. The pressure from alumni, leaders of political, cultural and social life of Wales was intense. It prompted Len Bigley to comment:

> In truth, the Blaenwern scheme at the very outset was an abject failure. The financial cutbacks were severe; the Agriculture Department had little or no interest in breeding cobs. To make matters worse, the legal documentation and covenants for the agreement were badly advised or remained unsigned. Enid and Pauline had unwisely used the services of the University solicitors.

Arrangements for the sale of the remaining 14 cobs went ahead, but this time they were not sold individually at open auction but unusually, as one lot in an attempt to keep the Llanarth stud bloodlines together under new ownership. They would be put out to tender and the highest bidder would win. The sale created worldwide interest and catalogues were sent to breeders in Europe, USA and Canada and the local press estimated the price would be around £20,000. Len Bigley had already confided to Anne Fowler that he would like to tender but he believed that he would not be allowed to do so as he had been a partner in the original business. The same was true of Anne Fowler who, like Len, remained a member of the trustees and, in addition, was legally involved in residual disputes. Apparently the College expected to make over £110,000 from the sale, an unrealistic figure and in addition they placed another stipulation on the tendering process, 'that preference would be given to a bid from someone living in Wales'.

This unexpected news provided the stimulus for Anne Fowler to mastermind a plan to salvage what remained of the Llanarth stud for Len Bigley. He would know how to restore the Llanarth breeding policy, his bid would be realistic and for Ann and Len it would, in some way, atone for the hurt and disappointment Enid and Pauline had endured. Anne Fowler had recently inherited part of her mother's estate and she felt that the money could be used to re-create the stud, but she and Len believed, perhaps wrongly, that, 'The college would not look kindly on a bid from either of them.' They decided to go through an intermediary. They approached a good friend of Pauline's, a well-known farmer and breeder, Ifan Phillips, who lived some three miles from Frongoch, in the Clarach valley. Anne Fowler put the plan to him. A bid in his own name would be submitted but of course in reality it would be on behalf of Len and Anne. Without hesitation he agreed. It was highly irregular, risky, if not illegal – secrecy was paramount. Anne Fowler reported the following short conversation during an interview for this book. 'Yes,' he said firmly, 'I'll put in a bid,' and Anne Fowler replied, 'And I'll put the money up front,' and without a pause Ifan Phillips said, 'I can trust you.' They both shook hands and Anne recalled twenty years later, 'Wasn't he wonderful! I thought the money could not have been spent in a better way than the re-creation of the stud.' As a final act, Anne Fowler decided to put in a bid in her own name, £2,000 lower than their bid through the intermediary, an attempt, she said, to put the college 'off the scent'.

Interest in the sale was increasing daily and on the first Tuesday of April 1985 bidding closed. Soon, it was announced that Ifan Phillips had put in the successful tender and, although it was not the highest, it did comply with one of the regulations. Mr Ifan Phillips lived in Wales and he immediately commented on his success, 'We purchased the Llanarth stud to perpetuate the breeding of Welsh cobs.'

The College had no idea that the real buyer lived in England. The cunning plan had worked but it was not divulged for a week or so. The facts were not revealed until a second HTV documentary was televised after the sale in 1985, an update on the sequence of events leading up to the final sale of the stud. The interviewer, Bruce Kent, asked the Principal, Gareth Owen, whether he knew the whereabouts of the Llanarth cobs. The Principal was nonplussed at such a question, but he replied that they were at a farm in the neighbouring valley, three miles away. 'No,' said the interviewer, 'they are a hundred miles away, in England, with Len Bigley.' Dr Owen's reaction could only reflect in words the surprise viewers could see on his face, 'I'm surprised, very surprised... astonished.' His astonishment was palpable for all to see, but I am not convinced over a quarter of a century later, that the manner in which the facts of the transaction were revealed 'on air' to the Principal and the public was ethical or necessary. At that time, the financial pressures on Dr Gareth Owen were intense and his efforts focused on saving core departments at the University College in a town where many would lose their jobs and where education was the main industry.

When Len Bigley left Blaenwern in 1982 to begin his own stud and livery business, he deliberately moved outside Wales to avoid any further acrimony with the College. They registered a new prefix in Ann's name, Tregarth, used for the horses they had bought and bred at their new stud, but when the cobs arrived from Wales they were able to transfer the Llanarth prefix. The Welsh cobs soon settled into their new home and as Ifan Phillips told Len and Anne Fowler when they had completed their private financial transaction to retain the Llanarth Cob bloodlines, 'The cobs are now back where they belong – with Len Bigley.'

But serious questions, mainly financial, continued to be asked of the College. Farmers were interested in finding out how the College accrued such losses in such a short space of time. Answers and financial details were hard to come by and even today, detailed audited accounts of those Blaenwern years have yet to emerge. They only show balance sheets and total losses. One fact is clear, the total loss for Blaenwern before the sale was over half a million pounds, a sum that horrified the average farmer. If a commercial farmer had run a business in this manner, he would quickly have been in great difficulty. This was a view which was regularly voiced during my research into the lives of Pauline and Enid and how their gift was executed and managed. How could a well-respected research and teaching Agriculture Department at the University College mismanage a gift of 380 acres and run up a debt of £500,000 in six years?

*   *   *   *

I was disappointed that the Director of Farms at the time, Dr David Morris, who was responsible for Blaenwern, refused to comment on any aspect or on any part of the Blaenwern story for this publication. My account is therefore one-sided and I have relied on published material and minutes of the UCW Council and the Finance and General Purposes Committee. The minutes tend to give resolutions of their deliberations and do not include a record of the discussions. Sadly, those who were directly responsible for taking executive policy decisions on key issues have died and I have resisted the temptation to depend on hearsay in order to obtain balance. The only interview given by Dr Morris to the media was televised in October 1983 in a television documentary after the College sale of the final group of horses. I repeat part of it here:

Bruce Kent: How did you get on with the old ladies of Llanarth?

Professor Morris: I got on very well with them. I spent many, many hours talking to them, and they were very reasonable people in many ways. They were old people and it was very difficult for them, you see. They lived in the house, the farm had been taken over, they had built up the stock over a number of years, and they regarded the farm as theirs. I would tell them how changes were made and they were very reasonable about it. I was very happy to tell them what was happening but sometimes they weren't happy although they were very reasonable people.

Bruce Kent: That's not the picture we have at all, Dr Morris. In fact I have a statement written by Miss Taylor after Miss Lewis died. She says, 'Miss Lewis who was a lady of gentle disposition and a kindly tolerant nature was driven by Dr Morris's rudeness to declare that she positively hated him and wouldn't see him at Blaenwern again.'

Professor Morris: Well really this is the first time I've heard of this, this very second. I was never rude to Miss Lewis. I got on very well with Miss Lewis. I don't know when this was written or why it was written. I didn't quarrel with her and we spent many hours in the house together talking. I set out to make money, to make ends meet at the farm. It was a difficult task. I don't know when this happened but it was a difficult situation for all of us. But never did Miss Lewis and I quarrel... not at all.

*   *   *   *

No one could really answer the question uppermost in everyone's mind. Would Pauline and Enid be satisfied in any way with the outcome? The question was put to the Registrar, Tom Owen, who had been involved in every negotiation. He answered, 'If it wasn't for the University, the stud would have disbanded five or six years ago, so we are very pleased that a large national enterprise had its life extended. I don't know whether the ladies would settle for that... maybe... I don't know.' Anne Fowler believed, 'Given the circumstances, I'm sure it's what Pauline would have wanted. The horses back with Len.' And Len Bigley commentated, 'You see Enid and Pauline had a vision. And all was lost before they died.'

POSTSCRIPT

I RETURNED TO BLAENWERN in February 2009 – the house appeared solid, stone-clad, resting, as if waiting patiently for the glory years to return. The sound of the classics on Radio 3 had ceased to resonate around the yard, the horseboxes were empty, and the stable doors were wide open. A horseshoe nailed to the stable door lintel was deeply scored with rust showing the seasonal passage of weather over the years. The cats and corgis have long since reached their own Valhalla and the cattle, sheep, and pigs, all Welsh breeds, have been removed from the map. The land remains part of the Cardiganshire hinterland farmed successfully by the Evans brothers.

Another Blaenwern farm situated 50 miles away in the magical hinterland of the Pembrokeshire Preseli hills provided inspiration for William Penfro Rowlands, another musician and teacher. He composed the memorable hymn tune 'Blaenwern' during the Welsh religious revival of 1904–5, in memory of the farm where he came to recuperate after a serious illness. At the time, he had moved from Pembrokeshire to Morriston in the Swansea valley where he taught music and conducted the Morriston United Choral Society. The tune became an instant favourite and today it is used regularly as an accompaniment to Charles Wesley's hymn, 'Love Divine All Love's Excelling'. Recently, a commemorative stone, commissioned by local residents, was unveiled near the Llys-y-Frân reservoir to mark William Rowland's achievements as a composer. His own birthplace, a cottage, Danycoed, is a ruin, a heap of stones, and they seem to reflect the economic downturn and depopulation of so many small farms in the uplands of Wales.

And it is to this area, the foothills of the Preseli hills, that Barbara Saunders Davies returned on her retirement in 1978 from her work at Emerson College, Forest Row to join other friends who had set up a 'bio-dynamic farming centre', Plas Dwbl, near Mynachlog-ddu, a village which is in the heart of the Preseli hills. She bought a small cottage, Llwynpiod, at the end of the drive to Plas Dwbl and there she deepened her knowledge of Rudolph Steiner and anthroposophy, and enjoyed working with visiting students who came to receive practical training in bio-dynamic farming. She was content and at peace back on her beloved land of Pembrokeshire with friends who had the same beliefs, taking an organic approach to sustainable agriculture. Plas Dwbl is a 100-acre farm of mainly grassland, some woodland and many small fields. Today, they have a herd of twenty Welsh Blacks and two Jersey cows and they grow vegetables on a small scale to sell at the local supermarket.

When Barbara first arrived to live in the community, 50 acres of land adjoining Plas Dwbl (at that time the farm was 50 acres) came on the market and she bought them to make it a more viable unit for their purpose. It is run by trustees of The Responsive Earth Trust which was merged in 1979 with the Living Earth Trust, a similar organisation established by Barbara at Emerson College as a result of her collaboration with Katherine Castelliz. They had formed a partnership translating German scientific books and a number of other projects. They compiled a series of Rudolph Steiner lectures, 'Nutrients and Stimulants', they wrote *Life on the land*, *The Plant* and *The Little Book of Compost* but their main occupation was the translation of Walter Cloos's *The Living Earth*.

I can find no reference, apart from the annual Christmas card and telephone calls, of meetings between Pauline, Enid and Barbara following Barbara's move to Pembrokeshire. Her friends at Plas Dwbl did not meet them, although they had heard many stories of her life at Pentre mansion, of her mother and brother, her niece Xina now living in France, experiences at Blaenwern and setting up the Llanarth stud. They heard very little of the sale and the last traumatic years. At Llwynpiod, in her old age Barbara wrote, 'They left the farm, the Llanarth stud and the herd of Welsh Blacks to the University at Aberystwyth in the hope that it would benefit the future of the breeds. The mismanagement and disasters that followed is common knowledge and was much publicised.'

But the cob world never forgot Barbara Saunders Davies. After leaving Blaenwern she often acted as judge, especially of palominos, at major shows and events and she never allowed her membership of the Welsh Pony and Cob Society to lapse. In 1991, she was 'elected to the role of Honorary Vice-President in recognition of conspicuous services rendered to the society and to the Welsh breeds.' Ten years later in 2001, she received a certificate from the society, 'As part of the centenary celebrations, the society wishes to acknowledge those who have been members for fifty years or more. In appreciation of the long-standing membership of Miss B Saunders Davies and in recognition of the valuable support given to the furtherance of the society and the Welsh breeds.'

When I visited Nim de Bruyne at Plas Dwbl who, with two friends (both called Mary) set up the centre, she described Barbara as a generous lady with a large circle of friends, and she added, 'She was so knowledgeable. She read so much. Somebody once said she was like a chest of drawers. You could open any drawer and Barbara knew something.' And she added, 'She loved to drive of course and this wasn't good when she was old and very arthritic, not good enough to drive... it wasn't right. So I went to the doctor and explained, and she never really forgave me for that.'

Barbara was ninety-six when she died. A team of carers and friends looked after her until the end. Phil Forder, in his address at the funeral, gave two enduring images of her:

The first would be Barbara in her chair by the window at Llwynpiod, cat on lap, surrounded by her books. Books were very important to her, on the table, on the floor, all open with bits of paper sticking out, all cross-referenced, books all around. Barbara was a learned scholar. The second picture is Barbara on the land. She had a deep love of nature and loved the farming life. I see her, stick in hand, cape on her shoulders, marching around the fields to inspect the cattle and hedges.

The Llanarth stud lives on and does so successfully at The Quakers Farm near Hay-on-Wye. Llanarth Nerissa, the mare Pauline gave Ann and Len Bigley as a wedding present thrived and at least a dozen or so of her progeny have become champions. Her last daughter, a brood mare aptly named Llanarth Swansong, became the champion part-bred at the Welsh Pony and Cob Society Show during the society's centenary year in 2002. But it is far from being a swansong for the stud. The record of Llanarth-bred horses at major shows around the country is formidable since they moved to Quakers Farm almost three decades ago. The stud now encompasses Welsh Section B ponies, part-breds and Ann's big hunter mares descended from Welsh part-breds. Their children, Simon and Catrin are also heavily involved in the enterprise. They have gleaned much information and knowledge from their parents on the intricacies of breeding – to breed for type, action, and above all for temperament. 'Put in time at home and practise to get them to perform without fear.'

Today Ann and Len are respected judges of cobs and ponies; they travel the world to events, 'I've been lucky, I've seen the world. My wife and I are off to Perth in Australia for two weeks to judge.' When I asked Len to name his favourite stallion, without hesitation he replied, 'Flying Comet. I suppose he is the one I achieved most with although Braint was marvellous.' Ann and Len have both played a major part in the affairs of the Welsh Pony and Cob Society in recent years and are considered to be wonderfully effective ambassadors and educators of the breed. Following a period as Chairman of the society, guiding the Council in its deliberations through a period of change and reform, Len was elected President in March 2008. 'It's the best equine society in the UK. Occasionally we get some internal wrangling, but the membership goes up and up, it is time now to look at the wider picture. It is a worldwide society of 8,500 members.'

When his term of office comes to an end in 2009, he will have accomplished a great deal, but he will have more time to revert to his first love, breeding cobs and ponies. Before my conversation with Len ended I asked him how he viewed the last chapter of events at Blaenwern. He paused, and his face suddenly expressed anxiety and distress, 'I suppose I should talk about it.' He paused, his wretchedness obvious to witness. 'You see, I was naïve and foolish... I don't think I would have cared as much if they had both died before the agreement to sell was taken.' And then he went on, 'When I consider the College expenditure, the huge debt and then to turn round and accuse Pauline of running a financial mess which was drowning her. That's what made me angry. It was simply not true.'

It was an unhappy ending to the lives of Pauline and Enid who had enriched so many aspects of life in Wales. Pauline said of Enid at Blaenwern, 'Enid loved every stick and stone of this place.' And Pauline prospered during her life – living her dream. But sadly in old age, their spirit had been broken.

The Ladies of Blaenwern is just one of a
whole range of publications from Y Lolfa.
For a full list of books currently in print,
send now for your free copy of our new
full-colour catalogue. Or simply
surf into our website

www.ylolfa.com

for secure on-line ordering.

TALYBONT CEREDIGION CYMRU SY24 5HE
e-mail ylolfa@ylolfa.com
website www.ylolfa.com
phone (01970) 832 304
fax 832 782